Understanding DC Circuits

Understanding DC Circuits

Dale R. Patrick

Stephen W. Fardo

Newnes

Boston Oxford Auckland Johannesburg Melbourne New Delhi

Newnes is an imprint of Butterworth–Heinemann.

Copyright © 2000 by Butterworth–Heinemann

℞ A member of the Reed Elsevier group

Library of Congress Cataloging-in-Publication Data

Patrick, Dale R.
    Understanding DC Circuits / by Dale R. Patrick and Stephen W. Fardo.
               p. cm.
    Includes index. (alk. pbk.)
    ISBN 0-7506-7110-6
    1. Electric circuits—Direct current. I. Fardo, Stephen W.
    II. Title.
    TK7816.F295 1999
    621.319′12—dc21   99-16983
                                                    CIP

British Library Cataloguing-in-Publication Data

A catalogue record for this book is available from the British Library.

The publisher offers special discounts on bulk orders of this book.

For information, please contact:
Manager of Special Sales
Butterworth-Heinemann
225 Wildwood Avenue
Woburn, MA 01801-2041
Tel: 781-904-2500
Fax: 781-904-2620

For information on all Butterworth–Heinemann publications available, contact our World Wide Web home page at: http://www.newnespress.com

Contents

UNIT THREE   **OHM'S LAW AND ELECTRIC CIRCUITS**

Preface

Understanding DC Circuits is an introductory text that provides coverage of the various topics in the field of direct current (dc) electronics. The key concepts presented in this book are discussed in a simplified approach that greatly enhances learning. The use of mathematics is discussed clearly through applications and illustrations.

Every unit is organized in a step-by-step progression of concepts and theory. Each unit begins with a *unit introduction* and *unit objectives*. A discussion of important *concepts* and *theories* follows. Numerous *self-examinations* with answers provided are integrated into each chapter to reinforce learning. *Experimental activities* with components and equipment listed are included with each unit to help students learn electronics through practical experimental applications. The final learning activity for each unit is a *unit examination*, which includes at least twenty objective, multiple-choice questions.

Definitions of *important terms* are presented at the beginning of each unit. Several *appendices* appear at the end of the book to aid students in performing experimental activities. The expense of the equipment required for the experiments is kept to a minimum. A comprehensive parts list is provided, as is information on electronics distributors.

The experiments suggested are low-cost activities that can be performed in the home or a school laboratory. They are very simple and easy to understand and emphasize troubleshooting concepts. The experiments allow students to develop an understanding of the topics discussed in each unit. They are intended as an important supplement to learning. Electronics can be learned experimentally at a low cost through completion of these labs. Appendices dealing with electronics symbols, safety, and soldering are provided for easy reference.

This textbook is organized in an easy-to-understand format. It can be used to acquire a basic understanding of electronics in the home, school, or workplace. The organization of the book allows students to progress at their own pace in the study of electronics. As students progress, they may wish to purchase various types of test equipment at varying degrees of expense.

Several *supplemental materials* are available to provide an aid to effective learning. These include the following:

1. *Instructor's Resource Manual*—provides the instructor with answers to all unit examinations and suggested data for experimental activities, including a comprehensive analysis of each experiment.

2. *Instructor's Transparency Masters*—enlarged reproductions of selected illustrations used in the textbook that are suitable for use for transparency preparation for class presentations.

3. *Instructor's Test Item File*—provides the instructor with many objective, multiple-choice questions for use with each unit of instruction.

These supplements are extremely valuable for instructors organizing electronics classes. The complete instructional cycle, from *objectives* to *evaluation*, is included in this book. We hope you will find *Understanding DC Circuits* easy to understand and that you are successful in your pursuit of knowledge in an exciting technical area. Electronics is an extremely vast and interesting field of study. This book provides a foundation for understanding electronics technology.

Dale R. Patrick
Stephen W. Fardo
Richmond, Kentucky

Course Objectives

Upon completion of this course on Understanding DC Circuits, you should be able to:

1. Understand the following basic electronics concepts:
   a. Voltage
   b. Current
   c. Resistance
   d. Power
   e. Static electricity
   f. Schematics

2. Use a multimeter to measure voltage, current, and resistance.

3. Solve basic electronics problems with dc circuits that involve calculation of voltage, current, and resistance.

4. Understand basic concepts of magnetism and electromagnetism.

5. Describe the construction, operation, and use of common electronics instruments.

6. Explain the properties of inductance and capacitance in dc circuits.

7. Construct experimental dc circuits using schematics and perform tests and measurements with a multimeter.

8. Understand basic safety rules and procedures involved in electronics applications.

9. Recognize common electronic components, symbols, and equipment.

10. Perform soldering operations to connect electronic components to circuit boards.

Parts List for Experiments

Various components and equipment are needed to perform the experimental activities in this course. These parts may be obtained from electronics suppliers, mail-order warehouses, or educational supply vendors. A list of several of these is included in appendix C.

These parts may be obtained through a variety of electronics suppliers. As a rule, a standard part number is used to obtain parts. In many cases, however, many other manufacturers make an equivalent part.

The following equipment and components are necessary for successful completion of the activities included in this book:

RESISTORS

    2 100 Ω, ½ W, 5% (brown-black-brown-gold)

    2 1000 Ω, ½ W, 5% (brown-black-red-gold)

    2 10 kΩ, ½ W, 5% (brown-black-orange-gold)

    1 1 MΩ, ½ W, 5% (brown-black-green-gold)

    1 1200 Ω, ½ W, 5% (brown-red-red-gold)

    1 200 Ω, ½ W, 5% (red-black-brown-gold)

    1 2000 Ω, ½ W, 5% (red-black-red-gold)

    1 22 kΩ, ½ W,5% (red-red-orange-gold)

    1 240 Ω, ½ W, 5% (red-yellow-brown-gold)

    1 300 Ω, ½ W, 5% (orange-black-brown-gold)

    1 390 Ω, ½ W, 5% (orange-white-brown-gold)

    1 510 Ω, ½ W, 5% (green-brown-brown-gold)

    1 750 Ω, ½ W, 5% (violet-green-brown-gold)

    1 910 Ω, ½ W, 5% (white-brown-brown-gold)

    1 470 Ω, ½ W, 5% (yellow-violet-brown-gold)

    1 68 Ω, ½ W, 5% (blue-gray-black-gold)

    1 68 kΩ, ½ W, 5% (blue-gray-orange-gold)

    1 100 Ω, 2 W, 5% (brown-black-brown-gold)

    2 100 Ω, 1 W, 5% (brown-black-brown-gold)

    1 15 Ω, ½ W, 5% (brown-green-black-gold)

    1 22 Ω , ½ W, 5% (red-red-black-gold)

    1 220 Ω, 2 W, 5% (red-red-brown-gold)

    1 2200 Ω, ¼ W, 5% (red-red-red-gold)

    1 220 kΩ, ½ W, 5% (red-red-yellow-gold)

    1 33 kΩ, ½ W, 5% (orange-orange-orange-gold)

    1 4700 Ω, ½ W, 5% (yellow-violet-red-gold)

    1 5100 Ω, ½ W, 5% (green-brown-red-gold)

CAPACITORS (ELECTROLYTIC)

1 10 μF, 35V

1 47 μF, 35V

MISCELLANEOUS

1 solder package

10 feet of no. 22 solid wire

1 compass

1 6 V light bulb

1 light socket

1 permanent magnet

1 small-parts container

1 6 V battery

2 1.5 V dry cells

1 200 Ω linear potentiometer

1 slide switch (DPDT)

1 relay

Basics of DC Electronics

Electronics is a fascinating science that we use in many different ways. It is difficult to count the many ways in which we use electronics each day. It is important for everyone today to understand electronics.

This unit deals with the most basic topics in the study of electronics. These include basic electric systems, energy and power, the structure of matter, electric charges, static electricity, electric current, voltage, and resistance. This unit and other units have definitions of important terms at the beginning. Preview these terms to gain a better understanding of what is discussed in the unit. As you study the unit, return to the definitions whenever the need arises. There are also self-examinations throughout the unit and a unit examination at the end of each unit. These will aid in understanding the material in the unit. Several experiments are suggested at the end of each unit. They may be completed in a laboratory or at home at low cost.

UNIT OBJECTIVES

Upon completing this unit, you will be able to do the following:

1. Explain the composition of matter.
2. Explain the laws of electric charges.
3. Define the terms *insulator, conductor,* and *semiconductor.*
4. Explain electric current flow.
5. Diagram a simple electric circuit.
6. Identify schematic electronic symbols.
7. Convert electric quantities from metric units to English units and English units to metric.
8. Use scientific notation to express numbers.
9. Define *voltage, current,* and *resistance.*
10. Describe basic types of batteries.

---

11. Connect batteries in series, parallel, and combination configurations.

12. Explain the purposes of different configurations of battery connections.

13. Explain factors that determine resistance.

14. Identify different types of resistors.

15. Identify resistor value according to color code and size.

16. Explain the operation of potentiometers (variable resistors).

17. Construct basic electronic circuits.

Important Terms

Before reading this unit, review the following terms. These terms provide a basic understanding of some of the concepts discussed. As you read other units, you may find it necessary to review these terms.

Ampere (A) The unit of electric charge, which is the basic unit of measurement for current flow in an electric circuit.

Atom The smallest particle to which an element can be reduced and still retain its characteristics.

Atomic number The number of particles called protons in the nucleus (center) of an atom.

Closed circuit A circuit that forms a complete path so that electric current can flow through it.

Compound The chemical combination of two or more elements to make an entirely different material.

Conductor A material that allows electric current to flow through it easily.

Control The part of an electric system that affects what the system does; a switch to turn on and turn off a light is a type of control.

Conventional current flow Current flow assumed to be in a direction from high charge concentration (+) to low charge concentration (−).

Coulomb (C) A unit of electric charge that represents a large number of electrons.

Current The movement of electric charge; the flow of electrons through an electric circuit.

Electromotive force (EMF) The pressure, or force, that causes electric current to flow.

Electron An atomic particle said to have a negative (–) electric charge; electrons are the means by which the transfer of electric energy takes place.

Electron current flow Current flow assumed to be in the direction of electron movement from a negative (–) potential to a positive (+) potential.

Electrostatic field The space around a charged material in which the influence of the electric charge is experienced.

Element The basic materials that make up all other materials; they exist by themselves (such as copper, hydrogen, carbon) or in combination with other elements (water is a combination of the elements hydrogen and oxygen).

Energy The capacity to do work.

Free electrons Electrons located in the outer orbit of an atom that are easily removed and result in flow of electric current.

Indicator The part of an electric system that shows whether the system is on or off or that a specific quantity is present.

Insulator A material that offers a high resistance to electric current flow.

Kinetic energy Energy that exists because of movement.

Load The part of an electric system that converts electric energy into another form of energy, such as an electric motor that converts electric energy into mechanical energy.

Matter Any material that makes up the world; anything that occupies space and has weight; a solid, a liquid, or a gas.

Metallic bonding The method by which loosely held atoms are bound together in metals.

Molecule The smallest particle to which a compound can be reduced before being broken down into its basic elements.

Neutron A particle in the nucleus (center) of an atom that has no electric charge, or is neutral.

Nucleus The core, or center part, of an atom; contains protons that have a positive charge and neutrons that have no electric charge.

Ohm (Ω) The unit of measurement of electric resistance.

Open circuit A circuit that has a broken path so that no electric current can flow through it.

Orbit The path along which electrons travel around the nucleus of an atom.

Orbital Areas through which electrons move; designated as s, p, d, and f.

Path The part of an electric system through which electrons travel from a source to a load, such as the electric wiring used in a building.

Potential energy Energy that exists because of position.

Power The rate at which work is done.

Proton A particle in the center of an atom that has a positive (+) electric charge.

Resistance *(R)* The opposition to the flow of electric current in a circuit; its unit of measurement is the ohm (Ω).

Semiconductor A material that has a value of electric resistance between that of a conductor and an insulator and is used to manufacture solid-state devices such as diodes and transistors.

Short circuit A circuit that forms a direct path across a voltage source so that a very high and possibly unsafe electric current flows.

Source The part of an electric system that supplies energy to other parts of the system, such as a battery that supplies energy for a flashlight.

Stable atom An atom that does not release electrons under normal conditions.

Static charge A charge on a material that is said to be either positive or negative.

Static electricity Electricity at rest caused by accumulation of either positive or negative electric charge.

Valence electrons Electrons in the outer orbit of an atom.

Volt (V) The unit of measurement of electric potential.

Voltage Electric force, or pressure, that causes current to flow in a circuit.

Watt (W) The unit of measurement of electric power.

Work The transforming or transferring of energy.

Electronic Systems

A simple electronic system block diagram and pictorial diagram are shown in Fig. 1-1. Using a block diagram allows a better understanding of electronic equipment and provides a simple way to "fit pieces together." The system block diagram can be used to simplify many types of electronic circuits and equipment.

The parts of an electronic system are the *source, path, control, load,* and *indicator.* The concept of electronic systems allows discussion of some complex things in a simplified manner. This method is used to present much of the material in this book to make it easier to understand.

(a)

(b)

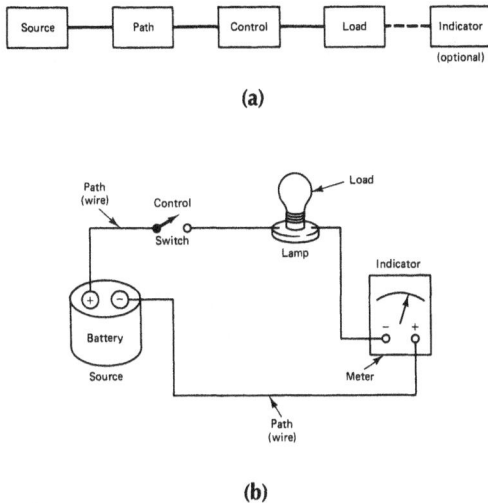

FIGURE 1-1 Electronic system. (a) Block diagram.
(b) Pictorial diagram.

The systems concept serves as a big picture in the study of electronics. In this way a system can be divided into a number of parts. The role played by each part then becomes clearer. It is easy to understand the operation of a complete electronic system. When one knows the function of a system, it is easier to understand the operation of each part. In this way, it is possible to see how the pieces of any electronic system fit together to make something that operates.

Each block of an electronic system has an important role to play in the operation of the system. Hundreds and even thousands of *components* sometimes are needed to form an electronic system. Regardless of the complexity of the system, each block must achieve its function when the system operates.

The *source* of an electronic system provides electric energy for the system. Heat, light, chemical, and mechanical energy may be used as sources of electric energy.

The *path* of an electronic system is simple compared with other system parts. This part of the system provides a path for the transfer of energy. It starts with the energy source and continues through the load. In some cases this path is a wire. In other systems a complex supply line is placed between the source and the load, and a return line from the load to the source is used. There usually are many paths within a complete electronic system.

The *control* section of an electronic system is the most complex part of the system. In its simplest form, control is achieved when a system is turned on or off. Control of this type takes place anywhere between the source and the load. The term *full control* is used to describe this operation. A system also may have some type of partial control. *Partial control* causes some type of operational change in the system other than turning it on or off. A change in the amount of current flow is a type of change achieved by means of partial control.

The *load* of an electronic system is the part or group of parts that do work. Work occurs when energy goes through a transformation or change. Heat, light, and mechanical motion are forms of work produced by loads. Much of the energy produced by the source is changed to another type by the load. The load usually is the most obvious part of the system because of the work it does. An example is a light bulb, which produces light.

The *indicator* of an electronic system displays a particular operating condition. In some systems the indicator is an optional part that is not really needed. In other systems it is necessary for proper operation. In some cases adjustments are made with indicators. In other cases an indicator is attached temporarily to the system to make measurements. Test lights, panel meters, oscilloscopes, and chart recorders are common indicators used in electronic systems.

Example of a System

Nearly everyone has used a flashlight. This device is designed to serve as a light source. A flashlight is a very simple type of electronic system. Figure 1-2 is a cutaway drawing of a flashlight with each part shown.

The battery of a flashlight serves as the energy *source* of the system. Chemical energy in the battery is changed into electric energy to cause the system to operate. The energy source of a flashlight—the battery—may be thrown away. Batteries are replaced periodically when they lose their ability to produce energy.

The *path* of a flashlight is a metal case or a small metal strip. Copper, brass, or plated steel is used as the path.

The *control* of electric energy in a flashlight is achieved by means of a slide switch or push-button switch. This type of control closes or opens the path between the source and the load device. Flashlights have only a means of full control, which is operated manually by a person.

The *load* of a flashlight is a small lamp bulb. When electric energy from the source passes through the lamp, the lamp produces a bright glow. Electric energy is changed into light energy. The lamp does a certain amount of work when this energy change takes place.

Flashlights do not have an *indicator* as part of the system. Operation is indicated, however, when the lamp produces light. The load of this system also acts as an indicator. In many electronic systems, the indicator is an optional part.

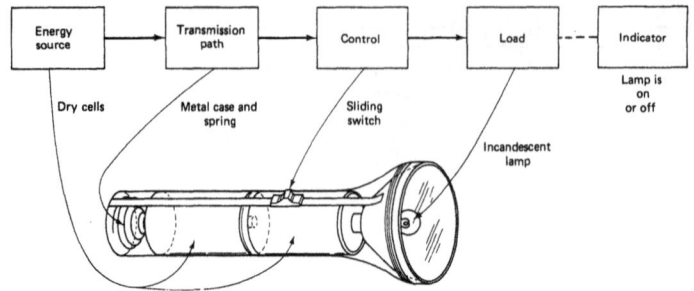

FIGURE 1-2 Cutaway drawing of a flashlight.

Energy, Work, and Power

An understanding of the terms *energy, work,* and *power* is necessary in the study of electronics. *Energy* means the capacity to do work. For example, the capacity to light a light bulb, to heat a home, or to move something requires energy. Energy exists in many forms, such as electric, mechanical, chemical, and thermal. If energy exists because of the movement of an object, such as a ball rolling down a hill, it is called *kinetic energy.* If it exists because the object is in position, such as a ball at the top of the hill but not yet rolling, it is called *potential energy.* Energy has become one of the most important factors in our society.

The second important term is *work.* Work is the transferring or transforming of energy. Work is done when a force is exerted to move something over a certain distance against opposition. Work is done when a chair is moved from one side of a room to the other. An electric motor used to drive a machine performs work. When force is applied to open a door, work is performed. Work is performed when motion is accomplished against the action of a force that tends to oppose the motion. Work also is done each time energy changes from one form into another.

The third important term is *power.* Power is the rate at which work is done. It involves not only the work performed but also the amount of time in which the work is done. For example, electric power is the rate at which work is done as electric current flows through a wire. Mechanical power is the rate at which work is done as an object is moved against opposition over a certain distance. Power is either the rate of production or the rate of use of energy. The *watt* (W) is the unit of measurement of power.

In the study of electronics, it is necessary to understand why electric energy exists. To gain this understanding, one must first look at how certain natural materials are made. It is then easier to see why electric energy exists.

Some basic scientific terms are often used in the study of chemistry. They also are important in the study of electronics. *Matter* is anything that occupies space and has weight. Matter can be a *solid*, a *liquid*, or a *gas*. Solid matter includes such things as metal and wood; liquid matter is exemplified by water or gasoline; and gaseous matter includes things such as oxygen and hydrogen. Solids can be converted into liquids, and liquids can be made into gases. For example, water can be a solid in the form of ice. Water also can be a gas in the form of steam. The difference is that the particles of which the substances are made move when heated. As they move, the particles strike one another and move farther apart. Ice is converted into a liquid by means of adding heat. If heated to a high temperature, water becomes a gas. All forms of matter exist in their most familiar forms because of the amount of heat they contain. Some materials require more heat than others to become liquids or gases. However, all materials can be made to change from a solid to a liquid or from a liquid to a gas if enough heat is added. These materials also can change into liquids or solids if heat is taken from them.

The next important term in the study of the structure of matter is *element*. An element is considered to be the basic material that makes up all matter. Materials such as hydrogen, aluminum, copper, iron, and iodine are a few of the more than 100 elements known to exist. A table of elements is shown in Fig. 1-3. Some elements exist in nature and some are manufactured. Everything around us is made of elements.

There are many more materials in our world than there are elements. Materials are made by means of combining elements. A combination of two or more elements is called a *compound*. For example, water is a compound made from the elements hydrogen and oxygen. Salt is made from sodium and chlorine.

Another important term is the *molecule*. A molecule is said to be the smallest particle to which a compound can be reduced before breaking down into its basic elements. For example, one molecule of water has two hydrogen atoms and one oxygen atom. Within each molecule one finds particles called *atoms*. Within these atoms are the forces that cause electric energy to exist. An atom is considered to be the smallest particle to which an element can be reduced and still have the properties of that element. If an atom were broken down any further, the element would no longer exist. The smallest particles in all atoms are called *electrons*, *protons*, and *neutrons*. Elements differ from one

Table of Selected Radioactive Isotopes

SARGENT-WELCH SCIENTIFIC COMPANY
7300 NORTH LINDER AVENUE SKOKIE, ILLINOIS 60077

Element	Symbol	Atomic no.	Atomic weight	Element	Symbol	Atomic no.	Atomic weight	Element	Symbol	Atomic no.	Atomic weight
Actinium	Ac	89	227*	Hafnium	Hf	72	178.6	Praseodymium	Pr	59	140.92
Aluminum	Al	13	26.97	Hahnium	Ha	105	262*	Promethium	Pm	61	145*
Americium	Am	95	243*	Helium	He	2	4.003	Protactinium	Pa	91	231*
Antimony	Sb	51	121.76	Holmium	Ho	67	164.94	Redium	Ra	88	226.05
Argon	Ar	18	39.944	Hydrogen	H	1	1.0080	Rodon	Rn	86	222

Element	Symbol	Atomic Number	Atomic Weight
Arsenic	As	33	74.91
Astatine	At	85	210*
Barium	Ba	56	137.36
Berkelium	Bk	97	247*
Beryllium	Be	4	9.013
Bismuth	Bi	83	209.00
Boron	B	5	10.82
Bromine	Br	35	79.916
Cadmium	Cd	48	112.41
Calcium	Ca	20	40.08
Californium	Cf	98	251*
Carbon	C	6	12.01
Cerium	Ce	58	140.13
Cesium	Cs	55	132.91
Chlorine	Cl	17	35.457
Chromium	Cr	24	52.01
Cobalt	Co	27	58.94
Copper	Cu	29	63.54
Curium	Cm	96	247
Dysprosium	Dy	66	162.46
Einsteinium	E	99	254*
Erbium	Er	68	167.2
Europium	Eu	63	152.0
Fermium	Fm	100	255*
Fluorine	F	9	19.00
Francium	Fr	87	233*
Gadolinium	Gd	64	156.9
Gallium	Ga	31	69.72
Germanium	Ge	32	72.60
Gold	Au	79	197.0
Indium	In	49	114.76
Iodine	I	53	126.91
Iridium	Ir	77	192.2
Iron	Fe	26	55.85
Krypton	Kr	36	83.8
Lanthanum	La	57	138.92
Lawrencium	Lw	103	257*
Lead	Pb	82	207.21
Lithium	Li	3	6.940
Lutetium	Lu	71	174.99
Magnesium	Mg	12	24.32
Manganese	Mn	25	54.94
Mendelevium	Mv	101	256*
Mercury	Hg	80	200.61
Molybdenum	Mo	42	95.95
Neodymium	Nd	60	144.27
Neon	Ne	10	20.183
Neptunium	Np	93	237*
Nickel	Ni	28	58.69
Niobium	Nb	41	92.91
Nitrogen	N	7	14.008
Nobelium	No	102	253
Osmium	Os	76	190.2
Oxygen	O	8	16.000
Palladium	Pd	46	106.7
Phosphorus	P	15	30.975
Platinum	Pt	78	195.23
Plutonium	Pu	94	244
Polonium	Po	84	210
Potassium	K	19	39.100
Rhenium	Re	75	186.31
Rhodium	Rh	45	102.91
Rubidium	Rb	37	85.48
Ruthenium	Ru	44	101.1
Rutherfordium or Kurchatonium	Rf or Ku	104	260*
Samarium	Sm	62	150.43
Scandium	Sc	21	44.96
Selenium	Se	34	78.96
Silicon	Si	14	28.09
Silver	Ag	47	107.880
Sodium	Na	11	22.997
Strontium	Sr	38	87.63
Sulfur	S	16	32.066
Tantalum	Ta	73	180.95
Technetium	Tc	43	97*
Tellurium	Te	52	127.61
Terbium	Tb	65	158.93
Thallium	Tl	81	204.39
Thorium	Th	90	232.12
Thulium	Tm	69	168.94
Tin	Sn	50	118.70
Titanium	Ti	22	47.90
Tungsten	W	74	183.92
Uranium	U	92	238.07
Vanadium	V	23	50.95
Xenon	Xe	54	131.3
Ytterbium	Yb	70	173.04
Yttrium	Y	39	88.92
Zinc	Zn	30	65.38
Zirconium	Zr	40	91.22

*Mass number of the longest-lived of the known available forms of the element.

Figure 1-3 Alphabetic list of the elements

another on the basis of the amounts of these particles in their atoms. The relations between matter, elements, compounds, molecules, atoms, electrons, protons, and neutrons is shown in Fig. 1-4.

The structure of the simplest atom, hydrogen, is shown in Fig. 1-5. The hydrogen atom has a center called a *nucleus,* which has one proton. A proton is a particle said to have a positive (+) charge. The hydrogen atom has one electron, which orbits around the nucleus of the atom. The electron is said to have a negative (−) charge. Most atoms also have neutrons in the nucleus. A neutron has neither a positive nor a negative charge and is considered neutral. The structure of a carbon atom is shown in Fig. 1-6. A carbon atom has six protons (+), six neutrons (N), and six electrons (−). The protons and the neutrons are in the nucleus, and the electrons *orbit* around the nucleus. The carbon atom has two orbits or circular paths. The first orbit contains two electrons. The other four electrons are in the second orbit.

Each atom has a different number of protons in the nucleus. This causes each element to have different characteristics. For example, hydrogen has one proton, carbon has six, oxygen has eight, and lead has 82. The number of protons that each atom has is called its *atomic number* (see Fig. 1-3).

The nucleus of an atom contains protons (+) and neutrons (N). Because neutrons have no charge and protons have positive charges, the nucleus of an atom has a positive charge. Protons are believed to be about one-third the diameter of electrons. The *mass* or weight of a proton is thought to be more than 1800 times greater than that of an electron. Electrons move easily in their orbits around the nucleus of an atom. It is the movement of electrons that causes electric energy to exist.

Early models of atoms showed electrons orbiting around the nucleus in analogy with planets around the sun. This model is inconsistent with much modern experimental evidence. Atomic *orbitals* are very different from the orbits of satellites.

Atoms consist of a dense, positively charged nucleus surrounded by a cloud or series of clouds of electrons that occupy energy levels, which are commonly called *shells.* The occupied shell of highest energy is known as the *valence shell,* and the electrons in it are known as *valence electrons.*

Electrons behave as both particles and waves, so descriptions of them always refer to their *probability* of being in a certain region around the nucleus. Representations of orbitals are boundary surfaces enclosing the probable areas in which the electrons are found. All *s* orbitals are spherical, *p* orbitals are egg shaped, *d* orbitals are dumbbell shaped, and *f* orbitals are double dumbbell shaped.

Covalent bonding involves *overlapping* of valence shell orbitals of different atoms. The electron charge becomes concentrated in this region and attracts the two positively charged nuclei toward the negative charge between them. In ionic bonding, the ions are discrete units. They group themselves in crystal structures, surrounding themselves with the ions of opposite charge.

The electrons of an atom occur in an exact pattern. The first orbit, or *shell,* contains up to 2 electrons. The next shell contains

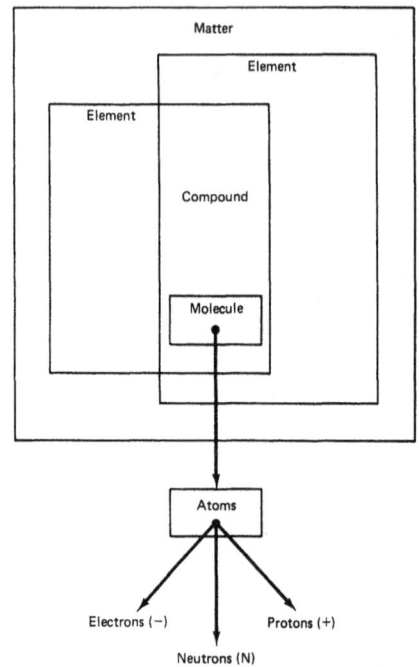

FIGURE 1-4 Structure of matter.

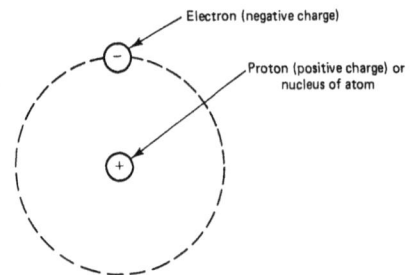

FIGURE 1-5 Structure of a hydrogen atom.

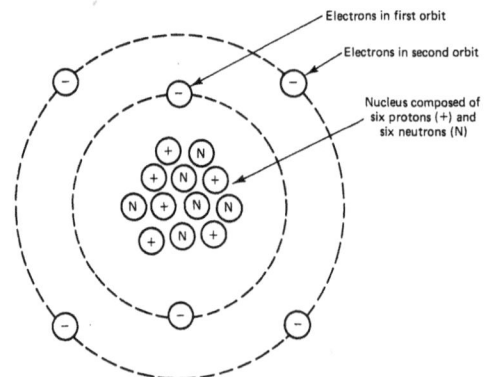

FIGURE 1-6 Structure of a carbon atom.

(a)

(b)

- Hydrogen atom
- Shared electron
- Oxygen atom
- Shared electron
- Hydrogen atom

(c)

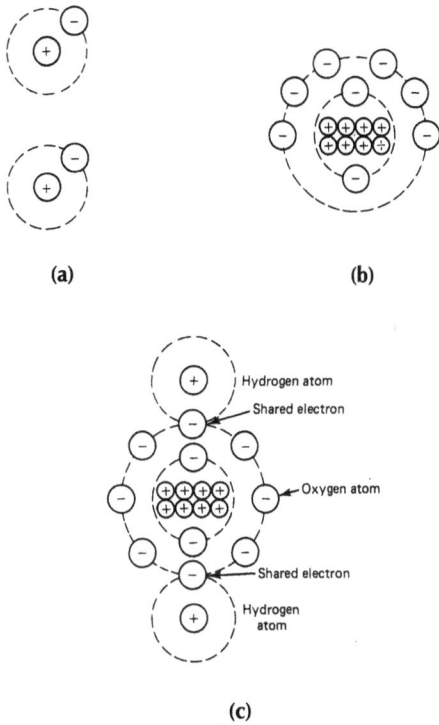

FIGURE 1-7 Water formed by combining hydrogen and oxygen. (a) Hydrogen atoms. (b) Oxygen atom. (c) Water molecule.

up to 8 electrons. The third contains up to 18 electrons. Eighteen is the largest quantity any shell can contain. New shells are started as soon as shells nearer the nucleus are filled with the maximum number of electrons.

Atoms with an incomplete outer shell are very *active*. When two unlike atoms with incomplete outer shells come together, they try to *share* their outer electrons. When their combined outer electrons are enough to make up one complete shell, *stable* atoms form. For example, oxygen has 8 electrons, 2 in the first shell and 6 in its outer shell. There is room for 8 electrons in the outer shell. Hydrogen has 1 electron in its outer shell. When two hydrogen atoms come near, oxygen combines with the hydrogen atoms by sharing the electrons of the two hydrogen atoms. Water is formed, as shown in Fig. 1-7. All the electrons are then bound tightly together, and a very stable water molecule is formed. The electrons in the incomplete outer shell of an atom are known as *valence electrons*. They are the only electrons that combine with other atoms to form compounds. They are also the only electrons used to cause electric current to flow. It is for this reason that it is necessary to understand the structure of matter.

Self-Examination

1. Energy caused by movement is called _____ energy.

2. Matter is anything that has _____ and occupies _____.

3. The basic material that makes up matter is the _____.

4. The smallest particle to which an element can be reduced is the _____.

5. The core of the atom is called the _____.

6. The nucleus of an atom contains neutral particles called _____.

7. Orbiting around the nucleus are one or more smaller particles of negative electric charge called _____.

8. An atom normally contains an equal number of electrons and protons and is said to be electrically _____.

9. Electrons on the outer shell of an atom are called _____.

10. An electronic system contains a(n) _____, _____, _____, _____, and _____.

Answers

1. Kinetic	2. Weight, space
3. Element	4. Atom
5. Nucleus	6. Neutrons
7. Electrons	8. Neutral
9. Valance electrons	10. Source, path, control, load, indicator

Electrostatic Charges

In the preceding section, the positive and negative charges of particles called protons and electrons are described. Protons and electrons are parts of atoms, which make up all things in our world. The positive charge of a proton is similar to the negative charge of an electron. However, a positive charge is the opposite of a negative charge. These charges are called *electrostatic charges*. Figure 1-8 shows how electrostatic charges affect one another. Each charged particle is surrounded by an *electrostatic field*.

The effect that electrostatic charges have on each other is important. They either repel (move away) or attract (come together) each other. This action is as follows:

1. Positive charges repel each other (Fig. 1-8a).

2. Negative charges repel each other (Fig. 1-8b).

3. Positive and negative charges attract each other (Fig. 1-8c).

Therefore it is said that *like charges repel* and *unlike charges attract*.

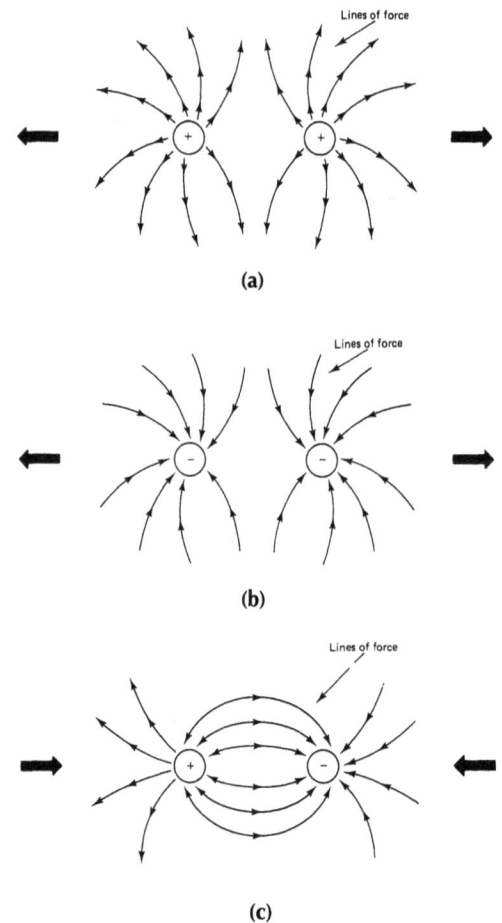

(a)

(b)

FIGURE 1-8 Electrostatic charges. (a) Positive charges repel. (b) Negative charges repel. (c) Positive and negative charges attract.

(c)

The atoms of some materials can be made to gain or lose electrons. The material then becomes charged. One way to do this is to rub a glass rod with a piece of silk cloth. The glass rod loses electrons (–), so it now has a positive (+) charge. The silk cloth pulls away electrons (–) from the glass. Because the silk cloth gains new electrons, it now has a negative (–) charge. Another way to charge a material is to rub a rubber rod with fur.

It is also possible to charge other materials. If a charged rubber rod is touched against another material, the new material may become charged. Some materials are charged when they are brought close to another charged object. Materials are charged because of the movement of electrons and protons. When an atom loses electrons (–), it becomes positive (+). These facts are very important in the study of electronics.

Charged materials affect each other because of lines of force. Try to visualize these as shown in Fig. 1-8. These imaginary lines cannot be seen; however, they exert a force in all directions around a charged material. Their force is similar to the force of gravity around the earth. This force is called a *gravitational field*.

Static Electricity

Most people have observed the effect of *static electricity*. Whenever objects become charged, it is the result of static electricity. A common example of static electricity is lightning. Lightning is caused by a difference in charge (+ and –) between the surface earth and clouds during a storm. The arc produced by lightning is the movement of charges between the earth and the clouds. Another common effect of static electricity is being shocked by touching a doorknob after walking across a carpeted floor. Static electricity also causes clothes taken from a dryer to cling together and hair to stick to a comb.

Electric charges are used to filter dust and soot in devices called *electrostatic filters*. Electrostatic precipitators are used in power plants to filter the exhaust gas that goes into the air. Static electricity also is used in the manufacture of sandpaper and the spray painting of automobiles. A device called an *electroscope* is used to detect a negative or positive charge.

Self-Examination

11. When electric charges exist it is called _____ electricity.

12. When a body of matter has more electrons than protons, it is said to have a _____ charge.

13. When a body of matter has more protons than electrons, it is said to have a _____ charge.

14. Like charges _____ and unlike charges _____.

15. Whenever two charged bodies are brought in contact, electrons move from the _____ charge to the _____ charge.

16. An application of static electricity is _____.

17. Materials are charged because of the _____ of electrons and protons.

18. The field about a charged body is generally represented by lines, which are referred to as electrostatic _____.

19. The lines of force are always shown leaving a _____ charge.

20. The lines of force are always shown entering a _____ charge.

Answers

11. Static	12. Negative
13. Positive	14. Repel, attract
15. Negative, positive	16. Electrostatic filters
17. Movement	18. Lines of force
19. Positive	20. Negative

Electric Current

Static electricity is caused by *stationary charges*. However, electric current is the *motion* of electric charges from one point to another. Electric current is produced when electrons (–) are removed from their atoms. Some electrons in the outer orbits of the atoms of certain elements are easy to remove. A force or pressure applied to a material causes electrons to be removed. The movement of electrons from one atom to another is called *electric current flow*.

Conductors

A material through which current flows is called a *conductor*. A conductor passes electric current easily. Copper and aluminum wire are commonly used as conductors. Conductors are said to have low *resistance* to electric current flow. Conductors usually have three or fewer electrons in the outer orbit of their atoms. Remember that the electrons of an atom orbit around the nucleus. Many metals are electric conductors. Each metal has a different ability to conduct electric current. For example, silver is a better conductor than copper, but silver is too expensive to use in large amounts. Aluminum does not conduct electric current as well as copper does. Aluminum is commonly used because it is less expensive and lighter than other conductors are. Copper is used more than any other conductor. Materials with only one outer orbit or valence electron (gold, silver, copper) are the best conductors.

Insulators

Some materials do not allow electric current to flow easily. The electrons of these *insulators* are difficult to release. The outer orbits of some insulators are filled with 8 electrons. In others, the outer orbits are more than half-filled with electrons. The atoms of materials that are insulators are said to be *stable*. Insulators have high resistance to the movement of electric current. Some examples of insulators are plastic and rubber.

Semiconductors

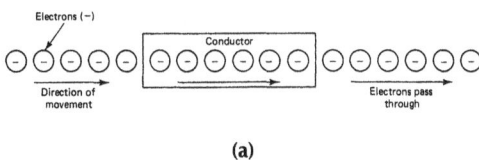

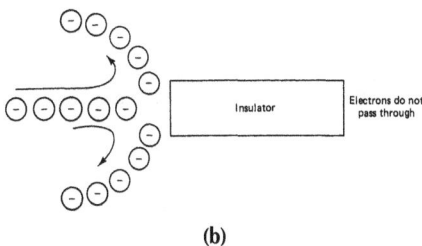

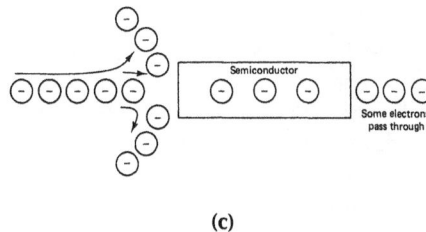

Materials called *semiconductors* have become very important in electronics. Semiconductor materials are not conductors and not insulators. They do not conduct electric current easily and are not good insulators. Their classification also depends on the number of electrons in the outer orbit of their atoms. Semiconductors have 4 electrons in their outer orbits. Remember that conductors

FIGURE 1-9 Comparison of conductors (a), insulators (b), and semiconductors (c).

have outer orbits *less than* half-filled and insulators have outer orbits *more than* half-filled. Figure 1-9 compares conductors, insulators, and semiconductors. Some common types of semiconductor materials are silicon, germanium, and selenium.

Current Flow

The usefulness of electricity is the result of what is called *electric current flow*. Current flow is the movement of electric charges along a conductor. Static electricity or electricity at rest has some practical uses because of electric charges. Electric current flow allows us to use electric energy to do many types of work.

The movement of outer-orbit electrons of conductors produces electric current. The electrons on the outer orbit of the atoms of a conductor are called *free electrons*. Energy released by these electrons as they move allows work to be done. As more electrons move along a conductor, more energy is released. This is called *increased* electric current flow. The movement of electrons along a conductor is shown in Fig. 1-10.

To understand how current flow takes place, it is necessary to know about the atoms of conductors. Conductors, such as copper, have atoms that are loosely held together. Copper is said to have atoms connected together by means of *metallic bonding*. A copper atom has one outer-orbit electron, which is loosely held to the atom. These atoms are so close together that their outer orbits overlap each other. Electrons can easily move from one atom to another. In any conductor the outer-orbit electrons constantly move in a random manner from atom to atom.

The random movement of electrons does not result in current flow. Electrons must move in the same direction to cause current flow. If electric charges are placed on each end of a conductor, the free electrons move in one direction. Figure 1-10 shows current flow through a conductor caused by negative (–) and positive (+) electric charges. Current flow takes place because the charges at each end of the conductor are different. Remember, like charges repel and unlike charges attract.

When an electric charge is placed on each end of the conductor, the free electrons move. Free electrons have a negative charge, so they are repelled by the negative charge on the left of Fig. 1-10. The free electrons are attracted to the positive charge on the right. The free electrons move to the right from one atom to another. If the charges on each end of the conductor increase, more free electrons will move. This increased movement causes more electric current flow.

Current flow is the result of electric energy produced as electrons change orbits. This impulse moves from one electron to another. When one electron (–) moves out of its orbit, it enters the orbit of another atom. An electron (–) is then repelled from that atom. This action goes on in all parts of a conductor. Remember that electric current flow is a transfer of energy.

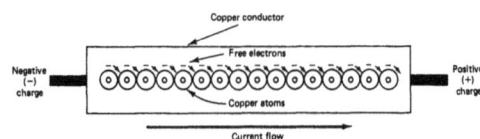

FIGURE 1-10 Current flow through a conductor.

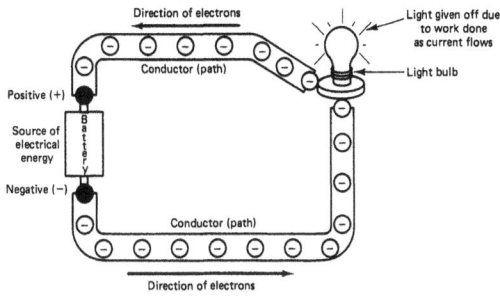

FIGURE 1-11 Current flow in a closed circuit.

Electronic Circuits

Current flow takes place in electronic circuits. A *circuit* is a path or conductor for electric current flow. Electric current flows only when it has a complete, or *closed-circuit,* path. There must be a source of electric energy to cause current to flow along a closed path. Figure 1-11 shows a battery used as an energy source to cause current to flow through a light bulb. The path, or circuit, is complete. Light is given off by the light bulb because of the work done as electric current flows through a closed circuit. Electric energy produced by the battery is changed to light energy in this circuit.

Electric current cannot flow if a circuit is open. An *open circuit* does not provide a complete path for current flow. If the circuit of Fig. 1-11 became open, no current would flow. The light bulb would not glow. Free electrons of the conductor would no longer move from one atom to another. An example of an open circuit occurs when a light bulb burns out. The filament (the part that produces light) opens. The opening in the filament stops current flow from the source of electric energy. This causes the bulb to stop burning, or producing light.

Another common circuit term is a *short circuit.* A short circuit can be very harmful. A short circuit occurs when a conductor connects directly across the terminals of an electric energy source. If a wire is placed across a battery, a short circuit occurs. For safety purposes, *a short circuit should never happen.* Short circuits cause too much current to flow from the source. The battery would probably be destroyed, and the wire might become hot or possibly melt because of a short circuit.

Direction of Current Flow

Electric current flow is the movement of electrons along a conductor. Electrons are negative charges. Negative charges are attracted to positive charges and repelled by other negative charges. Electrons move from the negative terminal of a battery to the positive terminal. This is called *electron current flow.* Electron current flow is in the direction of electron movement from negative to positive through a circuit.

Another way to look at electric current flow is in terms of charges. Electric charge movement is from an area of high charge to an area of low charge. A high charge can be considered positive and a low charge negative. With this method, an electric charge is considered to move from a high charge (positive) to a low charge (negative). This is called *conventional current flow.*

Electron and conventional current flow should not be confusing. They are two different ways of looking at current flow. One deals with electron movement and the other deals with charge movement. *In this book, electron current flow is used.*

Amount of Current Flow (the Ampere)

The amount of electric current that flows through a circuit depends on the number of electrons that pass a point in a certain time. The *coulomb* (C) is a unit of measurement of electric current. In electricity, many units of measurement are used. A coulomb is a large quantity of electrons. It is estimated that 1 C is 6,280,000,000,000,000,000 electrons (6.28×10^{18} in scientific notation). Because electrons are very small, it takes many to make one unit of measurement. When 1 C passes a point on a conductor in 1 second (s), 1 ampere (A) of current flows in the circuit. The unit is named for A.M. Ampere, an eighteenth-century scientist who studied electricity. Current is commonly measured in units called *milliamperes* (mA) and *microamperes* (μA). These are smaller units of current. A milliampere is 0.001 (1/1000) of an ampere, and a microampere is 0.000001 (1/1,000,000) of an ampere.

Current Flow Compared with Water Flow

An electric circuit is a path in which an electric current flows. Current flow is similar to the flow of water through a pipe. Electric current and water flow can be compared in some ways. Water flow is used to show how current flows in an electric circuit. When water flows in a pipe, something causes it to move. The pipe offers opposition or resistance to the flow of water. If the pipe is small, it is more difficult for the water to flow.

In an electric circuit, current flows through wires (conductors). The wires of an electric circuit are similar to the pipes through which water flows. If the wires are made of a material that has high resistance, it is difficult for current to flow. The result is the same as water flow through a pipe that has a rough surface. If the wires are large, it is easier for current to flow in an electric circuit. In the same way, it is easier for water to flow through a large pipe. Electric current and water flow are compared in Fig. 1-12. Current flows from one place to another in an electric circuit. Similarly, water that leaves a pump moves from one place to another. The rate of water flow through a pipe is measured in gallons per minute. In an electronic circuit, the current is measured in amperes. The flow of electric current is measured by the number of coulombs that pass a point on a conductor each second. A gallon of water is a certain number of molecules of water. A coulomb is a certain number of electrons. A current flow of 1 C/s makes 1A of current flow.

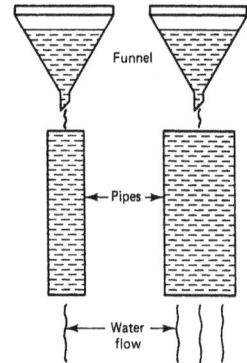

(a)

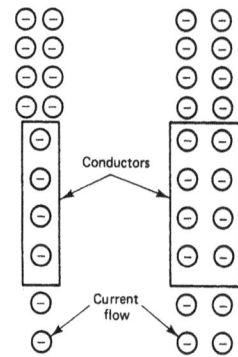

(b)

FIGURE 1-12 Comparison of electric current and water flow. (a) Water pipes. (b) Electric conductors.

Electric Force (Voltage)

Water pressure is needed to force water through a pipe. Similarly, electric pressure is needed to force current along a conductor. Water pressure usually is measured in pounds per square inch (lb/in²). Electric pressure is measured in volts (V). If a motor is rated at 120 V, 120 V of electric pressure must be applied to the motor to force the proper amount of current through it. More pressure would increase the current flow, and less pressure would not force enough current to flow. The motor would not operate properly with too much or too little voltage.

Water pressure produced with a pump causes water to flow through pipes. Pumps produce pressure, which causes water to flow. The same is true of an electric energy source. A source such as a battery or generator produces current flow through a circuit. As voltage is increased, the amount of current in a circuit also is increased. Voltage is also called *electromotive force (EMF)*. This term is largely responsible for the usage of E as an identifying letter for voltage. With the development of solid-state electronics the letter E has other meanings. To avoid duplications the letter V is now being used to identify voltage.

Resistance

The opposition to current flow in electric circuits is called *resistance*. Resistance is not the same for all materials. The number of free electrons in a material determines the amount of opposition to current flow. Atoms of some materials give up their free electrons easily. These materials offer low opposition to current flow. Other materials hold their outer electrons and offer high opposition to current flow.

Electric current is the movement of free electrons in a material. Electric current needs a source of electric pressure to move the free electrons through a material. Electric current does not flow if the source of electric pressure is removed. A material does not release electrons until enough force is applied. With a constant amount of electric force (voltage) and more opposition (resistance) to current flow, the number of electrons flowing (current) through the material is smaller. With constant voltage, current flow is increased by means of decreasing resistance. Decreased current results from more resistance. By increasing or decreasing the amount of resistance in a circuit, one can change the amount of current flow.

Materials that are good conductors have many free electrons. Insulating materials do not easily give up the electrons in the outer orbits of their atoms. Metals are the best conductors, copper, aluminum, and iron wire being the most common. Carbon and water are two nonmetal conductors. Materials such as glass, paper, rubber, ceramics, and plastics are common insulators.

Even very good conductors have some resistance, which limits the flow of electric current through them. The resistance of any material depends on the following four factors:

1. The material of which it is made

2. The length of the material

3. The cross-sectional area of the material

4. The temperature of the material

The material of which an object is made affects its resistance. The ease with which different materials give up their outer electrons is important in determining resistance. Silver is an excellent conductor of electricity. Copper, aluminum, and iron have more resistance but are more commonly used, because they are less expensive. All materials conduct an electric current to some extent, even though some (insulators) have very high resistance.

Length also affects the resistance of a conductor. The longer a conductor, the greater is the resistance. The shorter a conductor, the lower is the resistance. A material resists the flow of electrons because of the way in which each atom holds on to its outer electrons. The more material in the path of an electric current, the less current flow the circuit will have. If the length of a conductor is doubled, there is twice as much resistance in the circuit.

Another factor that affects resistance is the cross-sectional area of a material. The greater the cross-sectional area of a material, the lower is the resistance. The smaller this area, the higher is the resistance of the material. If two conductors have the same length but twice the cross-sectional area, the current flow is twice as much through the wire with the larger cross-sectional area. This happens because there is a wider path through which electric current can flow. Twice as many free electrons are available to allow current flow.

Temperature affects resistance. For most materials, at higher temperatures more resistance is offered to the flow of electric current. The colder the temperature, the less resistance a material offers to the flow of electric current. This effect is produced because a change in the temperature of a material changes the ease with which a material releases its outer electrons. A few materials, such as carbon, have lower resistance as the temperature increases. The effect of temperature on resistance varies with the type of material. The effect of temperature on resistance is the least important of the factors that affect resistance. A device called a *resistor*, which is used in electric circuits, is shown in Fig. 1-13.

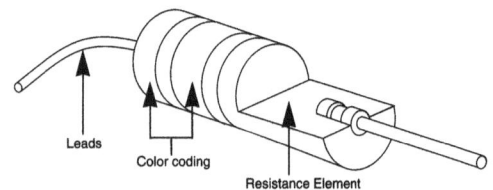

FIGURE 1-13 Resistor used in electric circuits.

Voltage, Current, and Resistance

We depend on electricity to do many things that sometimes are taken for granted. It is important to learn some of the basic electric terms commonly used in the study of electricity and electronics. The three basic electric terms are *voltage, current,* and *resistance.*

Voltage is best illustrated with a flashlight battery. The battery is a source of voltage. It is capable of supplying electric energy to a light connected to it. The voltage the battery supplies should be thought of as electric pressure. The battery has positive (+) and negative (−) terminals.

For a battery to supply electric pressure, a circuit must be formed. A simple electric circuit has a source, a conductor, and a load. An electric circuit is shown in Fig. 1-11. The battery is a source of electric pressure, or *voltage.* The conductor is a path that allows the electric current to pass the load. The lamp is called a load because it changes electric energy to light energy.

(a)

(b)

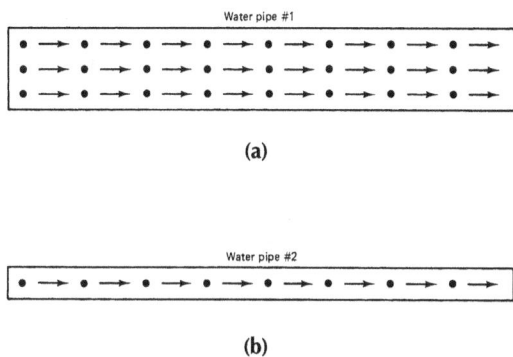

FIGURE 1-14 Water pipes showing the effect of resistance. (a) Many drops of water flow through water pipe 1. (b) Only a few drops of water flow through water pipe 2.

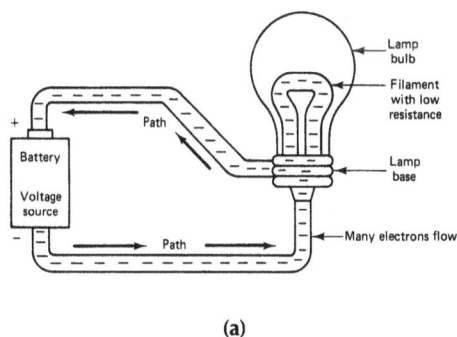

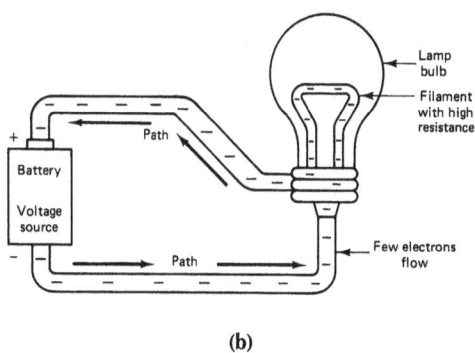

(a)

(b)

FIGURE 1-15 How lamp filament size affects current flow.

When the source, conductors, and load are connected together, a complete circuit is made. When the battery or voltage source is connected to the light bulb with conductors, current flows.

Current flows because of the electric pressure produced by the battery. The battery is similar to a water pump. A water pump also supplies pressure. Water in pipes is somewhat similar to the flow of current through a conductor. When the conductor is connected to the lamp, current flows. The current flow causes the lamp to light. Electric current is the flow of electrons through the conductor. Electrons move because of the pressure produced by the battery. Remember that electrons have a negative charge (–).

The movement of electrons through a conductor takes place at a rate based on the *resistance* of a circuit. A lamp offers resistance to the flow of electric current. Resistance is opposition to the flow of electric current. More resistance in a circuit causes less current to flow. Resistance can be explained with the example of two water pipes shown in Fig. 1-14. If a water pump is connected to a large pipe, such as pipe 1, water flows easily. The pipe offers a small amount of resistance to the flow of water. However, if the same water pump is connected to a small pipe, such as pipe 2, there is more opposition to the flow of water. The water flow through pipe 2 is less.

Inside a lamp bulb, the part that glows is called a *filament*. The filament is a wire that offers resistance to the flow of electric current. If the filament wire of a lamp is made of large wire, much current flows, as shown in the circuit of Fig. 1-15a. The filament offers a small amount of resistance to the flow of current. Figure 1-15b shows the circuit of a lamp with a filament of small wire. The small wire has more resistance, or opposition to current flow. Therefore less current flows in circuit b because it has higher resistance.

The terms *voltage, current,* and *resistance* are important. *Voltage* is electric pressure that causes current to flow in a circuit. *Current* is the movement of particles called electrons through a conductor in a circuit. *Resistance* is opposition to the flow of current in a circuit.

Volts, Ohms, and Amperes

There are many similarities between water systems and electric systems. These similarities help one to understand basic electric quantities. The volt (unit of electric pressure) is compared with the pressure that causes water to flow in pipes. Because the volt is a unit of electric pressure, it is always measured across two points. An electric pressure of 120 V exists across the terminals of electric outlets in the home. This value is measured with an electric instrument called a *voltmeter*.

The ampere, or amp, is a measure of the rate of flow of electric current or electron movement. Electric current is similar to the rate of flow of water in a pipe. Water flow is measured in gallons per minute. An ampere is a number of electrons per unit of time flowing in an electric circuit. An ampere is a measure of the rate of flow. An *ammeter* is used to measure the number of electrons that flow in a circuit.

When pressure is applied to a water pipe, water flows. The rate of flow is limited by friction in the pipe. When an electric pressure (voltage) is applied to an electric circuit, the resistance of the path limits the number of electrons (current) that flow. Resistance is measured with a meter called an *ohmmeter,* because the basic unit of resistance is the ohm.

Self-Examination

21. Substances that allow free motion of a large number of electrons are called _____.

22. Substances with tightly bound electrons are called _____.

23. Electric pressure is called _____.

24. The symbol for voltage is the letter _____.

25. The unit of measurement for voltage is the _____.

26. The flow of free electrons is called _____.

27. The unit of measurement for electric current is the _____.

28. Electron flow from negative to positive is called _____.

29. Opposition to current flow is called _____.

30. The unit of measurement for resistance is the _____.

Answers

21.	Conductors	22.	Insulators
23.	Voltage	24.	V
25.	Volt	26.	Current
27.	Ampere	28.	Electron current
29.	Resistance	30.	Ohm

Components, Symbols, and Diagrams

Most electronic equipment is made of several parts, or *components*, that work together. It would be almost impossible to explain how equipment operates without using symbols and diagrams. Electronic diagrams show how the component parts of equipment fit together. Common electronic components are easy to identify. It is also easy to learn the symbols used to represent electric components.

The components of electronic equipment work together to form a *system*. Anyone who studies electronics should be able to identify the components used in simple circuits. Components are represented by *symbols*. Symbols are used to make *diagrams*. A diagram shows how the components are connected together in a circuit. For example, it is easier to show symbols for a battery connected to a lamp than to draw a pictorial diagram of the battery and the lamp connected together. There are several symbols that you should learn to recognize. These symbols are used in many electronic diagrams. Diagrams are used for installing, troubleshooting, and repairing electronic equipment. The use of symbols makes it easy to draw diagrams and to understand the purpose of each circuit. Common electronic symbols are shown listed in appendix A.

In most electronic equipment wires (conductors) connect the components or parts to one another. The symbol for a conductor is a narrow line. If two conductors cross one another on a diagram, a symbol must be used to mark the point. Figure 1-16a shows two conductors that cross one another. If two conductors are connected together, a symbol is used to show the connection, as shown in Fig. 1-16b.

Figure 1-17 is a diagram of two lamps connected across a battery. The symbols for the battery and lamps are shown. Notice the part of the diagram where the conductors are connected together.

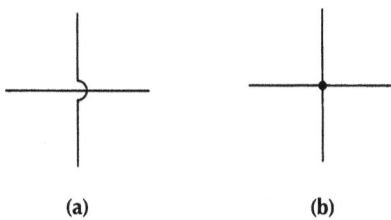

(a)   (b)

FIGURE 1-16 Symbols for electric conductors. (a) Conductors crossing. (b) Conductors connected.

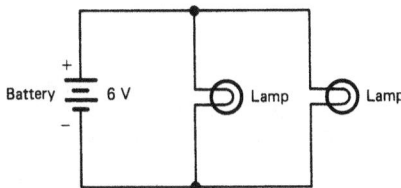

FIGURE 1-17 Symbols for a battery connected across two lamps.

A common electronic component is called a *switch*. Several types of switches are shown in Fig. 1-18. The simplest switch is a single-pole, single-throw (SPST) switch. This switch turns a circuit on or off. Fig. 1-19a shows the symbol for a switch in the *off*, or open, position. There is no path for current to flow from the battery to the lamp. The lamp is off when the switch is open. Figure 1-19b shows a switch in the *on*, or closed, position. This switch position completes the circuit and allows current to flow.

In many electronic circuits, a component called a *resistor* is used. Resistors are usually small, cylinder-shaped components. They are used to control the flow of electric current. A typical color-coded resistor and its symbol are shown in Fig. 1-20. The most common type of resistor has color coding to mark its value. Resistor values are always given in *ohms* (Ω). For example, a resistor might have a value of 100 Ω. Each color on the resistor represents a specific number. Resistor color-code values are easy to learn.

Another type of resistor is called a *potentiometer,* or *pot*. A pot is a variable resistor the value of which can be changed by means of adjustment of a rotary shaft. For example, a 1000 Ω pot can be adjusted to any resistance value from 0 to 1000 Ω by means of rotation of the shaft. The pictorial representation and symbol of this component are shown in Fig. 1-21a and b. In the example shown in Fig. 1-21c, potentiometer 1 is adjusted so that the resistance between points *A* and *B* is zero. The resistance between points *B* and *C* is 1000 Ω. When the shaft is turned as far in the opposite direction as it will go, the resistance between points *B* and *C* becomes zero (see potentiometer 2). Between points *A* and *B*, the resistance becomes 1000 Ω. When the shaft is rotated to the center of its movement, as shown by potentiometer 3, the resistance is split in half. The resistance from point *A* to point *B* is about 500 Ω and the resistance from point *B* to point *C* is also about 500 Ω.

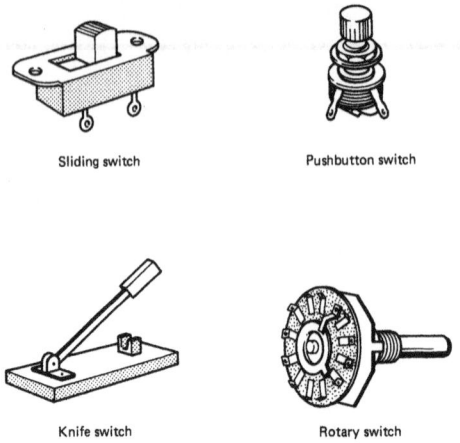

Sliding switch   Pushbutton switch

Knife switch   Rotary switch

FIGURE 1-18 Types of switches.

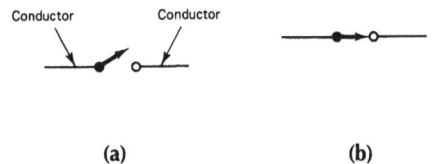

(a)   (b)

FIGURE 1-19 Symbol for a single-pole, single-throw (SPST) switch. (a) Open or off condition. (b) Closed or on condition.

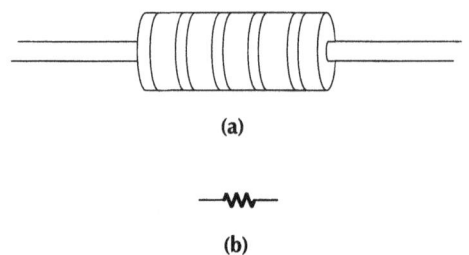

(a)

(b)

FIGURE 1-20 Resistors. (a) Color-coded resistor. (b) Symbol for a resistor.

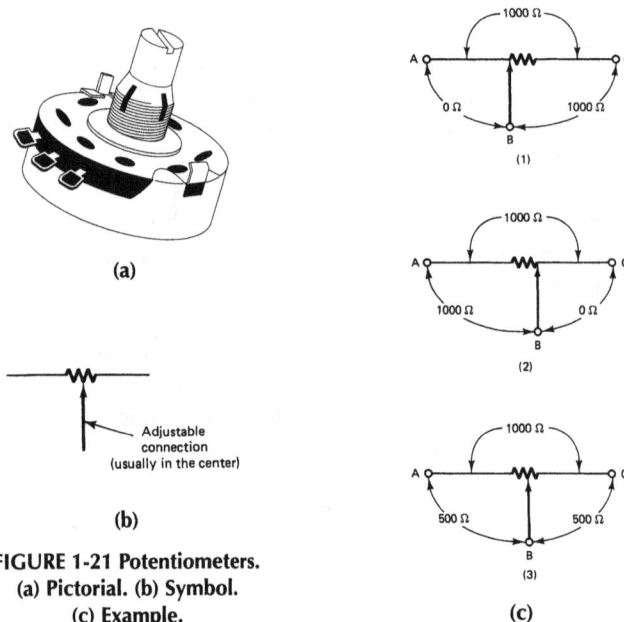

(a)

Adjustable connection (usually in the center)

(b)

**FIGURE 1-21 Potentiometers.
(a) Pictorial. (b) Symbol.
(c) Example.**

(1)

(2)

(3)

(c)

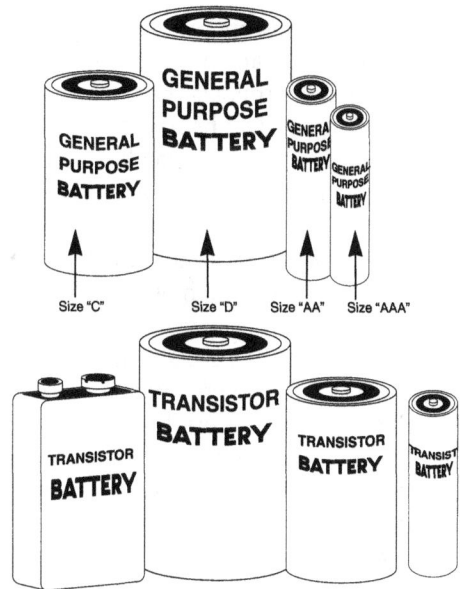

FIGURE 1-22 Common types of batteries.

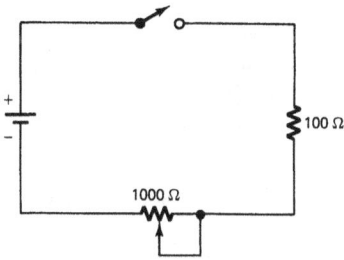

FIGURE 1-23 Simple circuit diagram.

The symbol for a battery is shown in Fig. 1-17. The symbol for any battery over 1.5 V is indicated by two sets of lines. A 1.5 V battery or cell is shown with one set of lines. The voltage of a battery is marked near its symbol. The long line in the symbol is always the positive (+) side, and the short line is the negative (−) side of the battery. Some common types of batteries are shown in Fig. 1-22

A simple circuit diagram with symbols is shown in Fig. 1-23. This diagram shows a 1.5 V battery connected to an SPST switch, a 100 Ω resistor, and a 1000 Ω potentiometer. Because symbols are used, no words have to be written beside them. Anyone using this diagram should recognize the components represented and how they fit together to form a circuit.

Resistors

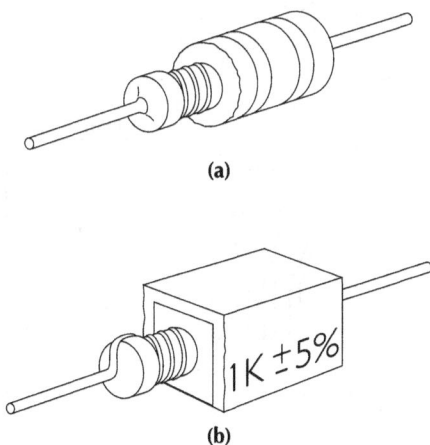

A wide variety of resistors are used. Resistors are made of either special carbon material or of metal film. Wire-wound resistors are ordinarily used to control large currents, and carbon resistors control currents that are smaller. Types of resistors are shown in Fig. 1-24.

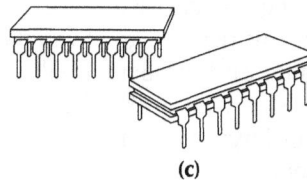

FIGURE 1-24 (a) Molded wire-wound resistor. (b) High-wattage wire-wound resistor. (c) 16-pin dual in-line package (DIP) resistor network used in printed circuit boards.

Wire-wound resistors are constructed by means of winding resistance wire on an insulating material. The wire ends are attached to metal terminals. An enamel coating is used to protect the wire and to conduct heat away from it. Wire-wound resistors may have fixed *taps,* which can be used to change the resistance value in steps. They may also have *sliders,* which can be adjusted to change the resistance to any fraction of their total resistance. Precision-wound resistors are used when the resistance value must be very accurate, such as in measuring instruments.

Carbon resistors are constructed of a small cylinder of compressed material. Wires are attached to each end of the cylinder. The cylinder is then covered with an insulating coating.

Variable resistors are used to change resistance while equipment is in operation. They are ordinarily called *potentiometers* or *rheostats.* Both carbon and wire-wound variable resistors are made on a circular form (Fig. 1-25a). A contact arm is attached to the form to make contact with the wire. The contact arm can be adjusted to any position on the circular form by means of a rotating shaft. A wire connected to the movable contact is used to vary the resistance from the contact arm to either of the two outer wires of the variable resistor. For controlling smaller currents, carbon variable resistors are made with a carbon compound mounted on a fiber disk (Fig. 1-25b). A contact on a movable arm varies the resistance as the arm is turned with a rotating metal shaft.

Resistor Color Codes

It is usually easy to find the value of a resistor from its color code or marked value. Most wire-wound resistors have resistance values (in ohms) printed on them. If they are not marked in this way, an ohmmeter must be used to measure the value. Most carbon resistors have colored bands to identify their value. A carbon resistor is shown in Fig. 1-26. This type has a color code.

Most resistors are color coded with an end-to-center color-band system of marking. In this color-coding system, three colors are used to indicate the resistance value in ohms. A fourth color is used to indicate the tolerance of the resistor. The colors are read in the correct order from the end of a resistor. Numbers from the resistor color code, shown in Fig. 1-27, are substituted for the colors. Through practice using the resistor color code, the value of a resistor may be determined at a glance.

It is difficult to manufacture a resistor to the exact value required. For many uses, the actual resistance value can be as much as 20% higher or lower than the value marked on the resistor without causing any problem. In most uses, the actual resistance does not have to be any closer than 10% higher or lower than the marked value. This percentage of variation between the marked color-code value and the actual value of a resistor is called *tolerance.* A resistor with a 5% tolerance should be no more than 5% higher or lower than the marked value. Resistors with tolerances of lower than 5% are called *precision resistors.*

Resistors usually are marked with color bands at one end. For example, a resistor may have three color bands (yellow, violet, and brown) at one end. The color bands are read from the end

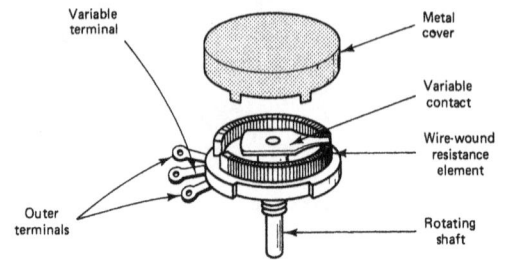

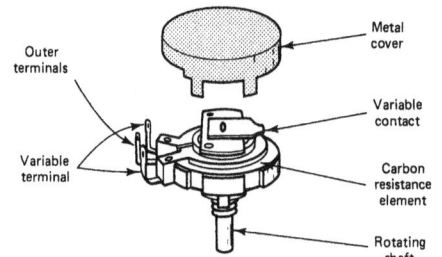

FIGURE 1-25 Variable resistor construction. (a) Wire-wound variable resistor. (b) Carbon variable resistor.

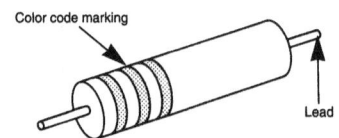

FIGURE 1-26 Carbon resistor.

Color	1st digit	2nd digit	Number of zeros or multiplier	Tolerance (%)
Black	0	0	1	
Brown	1	1	10	
Red	2	2	100	
Orange	3	3	1,000	
Yellow	4	4	10,000	
Green	5	5	100,000	
Blue	6	6	1,000,000	
Violet	7	7		
Gray	8	8		
White	9	9		
Gold*	–	–	0.1	5
Silver*	–	–	0.01	10
No color	–	–		20

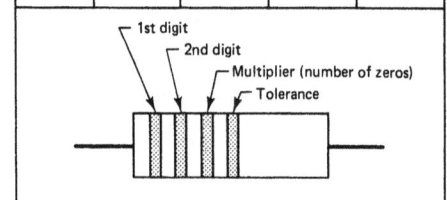

*When resistors have a value of less than 10 Ω, the third color band is a decimal multiplier. The two colors used are: gold = × 0.1 and silver = × 0.01.

FIGURE 1-27 Resistor color code.

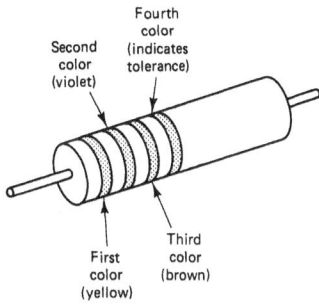

FIGURE 1-28 End-to-center system for color-code marking of resistors. Resistor value = 470 Ω.

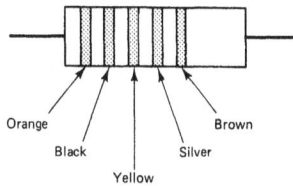

Resistor value = 300,000 Ω; 10% tolerance

FIGURE 1-29 Example of resistor color code.

toward the center, as shown in Fig. 1-28. The resistance value is 470 Ω. When black is the center dot or third color no zeros are to be added to the digits. A resistor with a green band, a red band, and a black band has a value of 52 Ω.

Another example of a resistor with the color-code marking system is shown in Fig. 1-29. Read the colors from left to right from the end of the resistor where the bands begin. Use the resistor color chart in Fig. 1-27 to determine the value of the resistor in ohms and its tolerance.

The first color is orange, so the first digit in the value of the resistor is 3. The second color is black, so the second digit in the value of the resistor is 0. The first two digits of the resistor value are 30. The third color is yellow, which indicates the number by which the first two digits are to be multiplied. Sometimes it is easier to think of the third color as the number of zeros to add to the first two digits. The color yellow is a multiplier of 10,000.

Multiply 30 by 10,000 to obtain the value of the resistor (30 × 10,000 = 300,000 Ω). Yellow means that four zeros are added to the first two digits to find the value of the resistor (30 with four zeros added equals 300,000 Ω).

The fourth color is silver. The tolerance of the resistor is given with this band. The tolerance shows how close the actual value of the resistor should be to the color-code value. Tolerance is a percentage of the actual value. In this example, silver shows a tolerance of ±10%. This means that the resistor should be within 10% of 300,000 Ω in either direction. The value of the resistor could be as low as 270,000 Ω (300,000 × 0.10 = 30,000 and 300,000 − 30,000 = 270,000). The value could be as high as 330,000 Ω (300,000 × 0.10 = 30,000 and 300,000 + 30,000 = 330,000).

Sometimes there is no fourth color on the body of the resistor. Then the tolerance is 20%. With the color code, it is easy to list the ohm value and tolerance of resistors. Standard values of 5% and 10% tolerance color-coded resistors are listed in Fig. 1-30.

					10% tolerance						
0.27	1.2	5.6	27	120	560	2700	12,000	56,000	0.27M	1.2M	5.6M
0.33	1.5	6.8	33	150	680	3300	15,000	68,000	0.33M	1.5M	6.8M
0.39	1.8	8.2	39	180	820	3900	18,000	82,000	0.39M	1.8M	8.2M
0.47	2.2	10	47	220	1000	4700	22,000	0.1M	0.47M	2.2M	10M
0.56	2.7	12	56	270	1200	5600	27,000	0.12M	0.56M	2.7M	12M
0.68	3.3	15	68	330	1500	6800	33,000	0.15M	0.68M	3.3M	15M
0.82	3.9	18	82	390	1800	8200	39,000	0.18M	0.82M	3.9M	18M
1.0	4.7	22	100	470	2200	10,000	47,000	0.22M	1.0M	4.7M	22M

					5% tolerance						
0.24	1.1	5.1	24	110	510	2400	11,000	51,000	0.24M	1.1M	5.1M
0.27	1.2	5.6	27	120	560	2700	12,000	56,000	0.27M	1.2M	5.6M
0.30	1.3	6.2	30	130	620	3000	13,000	62,000	0.30M	1.3M	6.2M
0.33	1.5	6.8	33	150	680	3300	15,000	68,000	0.33M	1.5M	6.8M
0.36	1.6	7.5	36	160	750	3600	16,000	75,000	0.36M	1.6M	7.5M
0.39	1.8	8.2	39	180	820	3900	18,000	82,000	0.39M	1.8M	8.2M
0.43	2.0	9.1	43	200	910	4300	20,000	91,000	0.43M	2.0M	9.1M
0.47	2.2	10	47	220	1000	4700	22,000	0.1M	0.47M	2.2M	10M
0.51	2.4	11	51	240	1100	5100	24,000	0.11M	0.51M	2.4M	11M
0.56	2.7	12	56	270	1200	5600	27,000	0.12M	0.56M	2.7M	12M
0.62	3.0	13	62	300	1300	6200	30,000	0.13M	0.62M	3.0M	13M
0.68	3.3	15	68	330	1500	6800	33,000	0.15M	0.68M	3.3M	15M
0.75	3.6	16	75	360	1600	7500	36,000	0.16M	0.75M	3.6M	16M
0.82	3.9	18	82	390	1800	8200	39,000	0.18M	0.82M	3.9M	18M
0.91	4.3	20	91	430	2000	9100	43,000	0.20M	0.91M	4.3M	20M
1.0	4.7	22	100	470	2200	10,000	47,000	0.22M	1.0M	4.7M	22M

FIGURE 1-30 Standard values of color-coded resistors (Ω).

Another way to determine the value of color-coded resistors is to remember the following:

Big Brown Rabbits Often Yield Great Big Vocal Groans When Gingerly Slapped.

The first letter of each word in the quotation is the same as the first letter in each of the colors used in the color code. The words of the quote are counted (beginning with zero) to find the word corresponding to each digit or the number of zeros to be added (Fig. 1-31). A method such as this quotation should be used to remember the color code.

Big	Brown	Rabbits	Often	Yield	Great	Big	Vocal	Groans	When	Gingerly	Slapped
B	B	R	O	Y	G	B	V	G	W	G	S
l	r	e	r	e	r	l	i	r	h	o	i
a	o	d	a	l	e	u	o	a	i	l	l
c	w		n	l	e	e	l	y	t	d	v
k	n		g	o	n		e		e		e
			e	w			t				r
↓	↓	↓	↓	↓	↓	↓	↓	↓	↓	↓	↓
0	1	2	3	4	5	6	7	8	9	5%	10%

FIGURE 1-31 Quotation used to memorize resistor color code.

Power Rating of Resistors

The size of a resistor helps to determine the power rating. Larger resistors are used in circuits that have high power ratings. Small resistors will be damaged if they are put in high-power circuits. The power rating of a resistor indicates its ability to give off or dissipate heat. Common power (wattage) ratings of color-coded resistors are ⅛, ¼, ½, 1, and 2 W. Resistors that are larger in physical size will give off more heat and have higher power ratings.

Schematics

Schematics are used to represent the parts of electronic equipment or circuits. They show how the components or parts of each circuit fit together. Schematics are used to show the details of the electric connections of any type of circuit or system. Schematics are used by manufacturers of electronic equipment to show operation and as an aid in servicing the equipment. A typical schematic is shown in Fig. 1-32. Symbols are used to represent electronic components in schematics. All equipment manufacturers use standard electronics symbols. Some common basic electronics symbols are shown in appendix A. These symbols should be memorized.

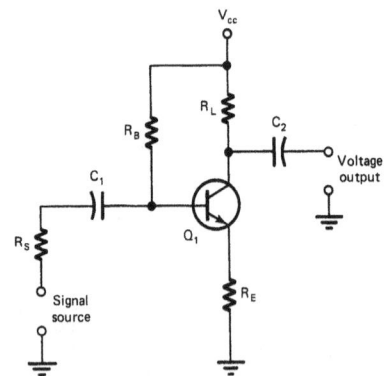

FIGURE 1-32 Schematic of a transistor amplifier circuit.

Block Diagrams

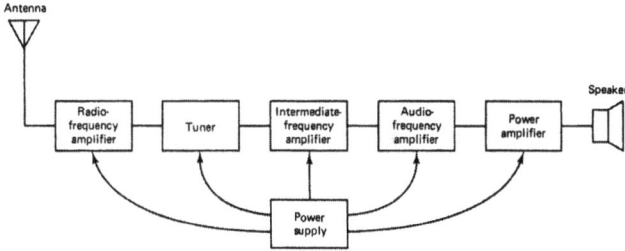

FIGURE 1-33 Block diagram of the parts of a radio.

Another way to show how electronic equipment operates is to use *block diagrams.* Block diagrams show the functions of the subparts of any electronic system. A block diagram of an electronic system is shown in Fig. 1-1a. The same type of diagram is used to show the parts of a radio in Fig. 1-33. Inside the blocks, symbols or words are used to describe the function of the block. Block diagrams usually show the operation of the entire system. They provide an idea of how a system operates. However, they do not show detail the way a schematic diagram does. It is easy to see the main subparts of a system by looking at a block diagram.

Wiring Diagrams

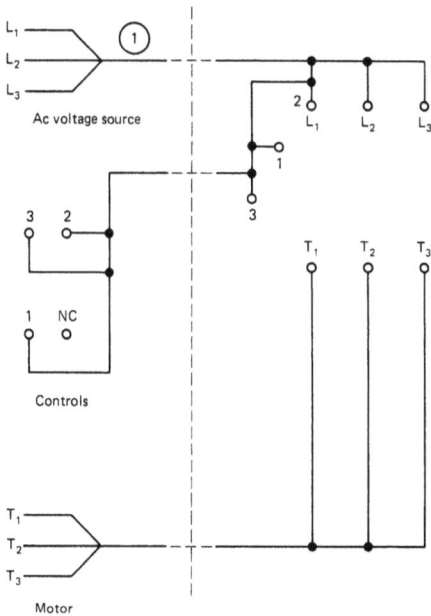

FIGURE 1-34 Simple wiring diagram.

Another type of electronic diagram is a *wiring diagram* (sometimes called a *cabling diagram*). Wiring diagrams show the actual location of parts and wires on equipment. The details of each connection are shown on a wiring diagram. Schematics and block diagrams show only how parts fit together electrically. Wiring diagrams show the details of actual connections. A simple wiring diagram is shown in Fig. 1-34.

Self-Examination

Draw the symbols for the following components:

31. Resistor

32. SPST switch

33. Lamp

34. Ground connection

35. Potentiometer

36. Fuse

Using Fig. 1-28, write the resistance value and tolerance of each color-coded resistor.

37. (1) Violet
    (2) Green
    (3) Orange
    (4) Gold
    _____ Ohms
    _____ Tolerance

38. (1) Yellow
    (2) Violet
    (3) Green
    _____ Ohms
    _____ Tolerance

39. (1) Green
    (2) Blue
    (3) Red
    _____ Ohms
    _____ Tolerance

40. (1) Red
    (2) Red
    (3) Blue
    _____ Ohms
    _____ Tolerance

41. (1) Brown
    (2) Black
    (3) Brown
    (4) Gold
    _____ Ohms
    _____ Tolerance

42. (1) Gray
    (2) Red
    (3) Black
    (4) Silver
    _____ Ohms
    _____ Tolerance

43. (1) White
    (2) Brown
    (3) Orange
    (4) Gold
    _____ Ohms
    _____ Tolerance

44. (1) Yellow
    (2) Violet
    (3) Orange
    _____ Ohms
    _____ Tolerance

45. (1) Brown
    (2) Green
    (3) Brown
    (4) Gold
    _____ Ohms
    _____ Tolerance

31. —ᴧᴧ—	32. —o⟋ o—	
33. (image)	34. (ground symbol)	
35. —ᴧᴧᴧ— (arrow)	36. —o◠o—	

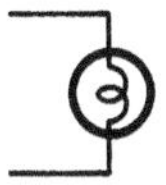

37. 75,000 ± 5%	38. 4,700,000 ± 20%
39. 5,600 ± 20%	40. 22,000,000 ± 20%
41. 100 ± 5%	42. 82 ± 10%
43. 91,000 ± 5%	44. 47,000 ± 20%
45. 150 ± 5%	

Electric Units

All quantities can be measured. The distance between two points may be measured in meters, kilometers, inches, feet, or miles. The weight of an object may be measured in ounces, pounds, grams, or kilograms. Electric quantities may also be measured. The more common electric units of measurement are discussed next.

There are four common units of electric measurement. *Voltage* is used to indicate the force that causes electron movement, or *current*. *Resistance* is the opposition to current (electron movement). The amount of work done or energy used in the movement of electrons in a given period of time is called *power*. Figure 1-35 shows the four basic units of electric measurement.

Electrical quantity	Unit of measurement	Symbol	Description
Voltage	Volt (V)	V^* or E	Electrical "pressure" that causes current flow
Current	Ampere (A)	I	Amount of electron movement through a circuit
Resistance	Ohm (Ω)	R	Opposition to current flow
Power	Watt (W)	P	Amount of work done as current flows through a circuit

*V is used in this book.

FIGURE 1-35 Four basic units of electric measurement.

Small Units

The electric unit used to measure a certain value often is less than a whole unit (less than 1). Examples of this are 0.6 V, 0.025 A, and 0.0550 W. When this occurs, prefixes are used. Some prefixes are shown in Fig. 1-36.

For example, a millivolt is 1/1000 of a volt, and a microampere is 1/1,000,000 of an ampere. The prefixes in Fig. 1-36 may be used with any electric unit of measurement. The unit is divided by the fractional part of the unit. For example, if 0.6 V is to be changed to millivolts, 0.6 V is divided by the fractional part of the unit. So 0.6 V equals 600 (mV), because 0.6 ÷ 0.001 = 600 mV. Also, 0.0005 A equals 500 μA. When changing a basic electric unit to a prefix unit, move the decimal point of the unit to the right by the same number of places in the fractional prefix. To change 0.8 V to millivolts, move the decimal point of 0.8 V three places to the right (800), because the prefix *milli* has three decimal places. So 0.8 V equals 800 mV. The same method is used for converting any electric unit to a unit with a smaller prefix.

An electric unit with a prefix often is converted to the basic unit. For example, milliamperes may be converted to amperes. Microamperes sometimes are converted to amperes. Microvolts sometimes are converted to volts. When a unit with a prefix is converted to a basic unit, the prefix must be multiplied by the fractional part of the whole unit of the prefix. For example, 68 mV converted to volts equals 0.068 V (68 mV × 0.001 = 0.068 V).

When changing a fractional prefix unit into a basic electric unit, move the decimal in the prefix unit to the left by the same number of places of the prefix. To change 225 mV to volts, move the decimal point in 225 three places to the left (225) because the prefix milli- has three decimal places. So, 225 mV equals 0.225 V. The same method is used when changing any fractional prefix unit back to the basic electric unit.

Prefix	Abbreviation	Fractional part of a whole unit
milli	m	1/1000 or 0.001 (3 decimal places)
micro	μ	1/1,000,000 or 0.000001 (6 decimal places)
nano	n	1/1,000,000,000 or 0.000000001 (9 decimal places)
pico	p	1/1,000,000,000,000 or 0.000000000001 (12 decimal places)

FIGURE 1-36 Prefixes of units smaller than one.

Large Units

Electric measurements such as 20,000,000 W, 50,000 Ω, and 38,000 V are very large. When this occurs, prefixes are needed to make these large numbers easier to use. Some prefixes used for large electric values are shown in Fig. 1-37. To change a large value to a smaller unit, divide the large value by the number of the prefix. For example, 48,000,000 Ω is changed to 48 MΩ by means of dividing 48,000,000 by 1,000,000. To convert 7000 V to kilovolts (kV), divide 7000 by 1000 (7000 ÷ 1000 = 7 kV). To change a large value to a prefix unit, move the decimal point in the large value to the left by the number of zeros of the prefix. Thus 3600 V equals 3.6 kV (3600). To convert a prefix unit to a larger unit, the decimal point is moved to the right by the same number of places in the unit. The number also may be multiplied by the value of the prefix. If 90 MΩ is converted to ohms, the decimal point is moved six places to the right (90,000,000). The 90 MΩ value also may be multiplied by the value of the prefix, which is 1,000,000. Thus 90 MΩ × 1,000,000 = 90,000,000Ω.

The simple conversion scale shown in Fig. 1-38 is useful when converting large and small units to units of measurement with prefixes. This scale uses either powers of 10 or decimals to express the units.

Prefix	Abbreviation	Number of times larger than 1
Kilo	k	1000
Mega	M	1,000,000
Giga	G	1,000,000,000

FIGURE 1-37 Prefixes of large units.

CONVERSION SCALE FOR LARGE OR SMALL NUMBERS

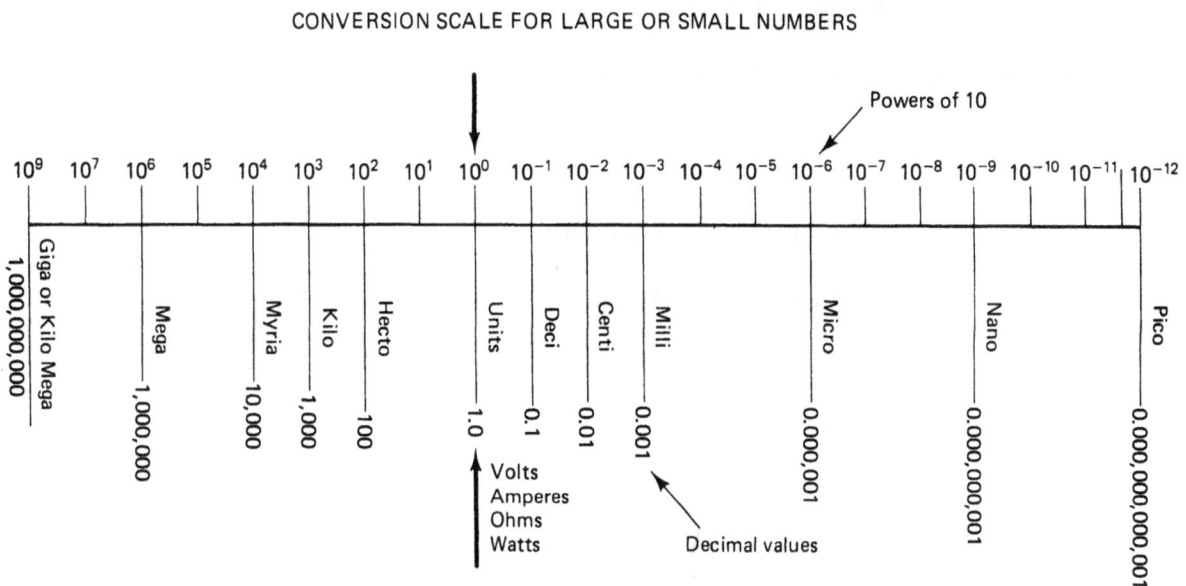

Directions for using the conversion scale:

1. Find the position of the term as expressed in its original form. ➞ 20 μA
2. Select the position of the conversion unit on the scale. ➞ Amperes
3. Write the original number as a whole number or in powers of 10. ➞ 20 μA or 20 × 10^{-6} A
4. Shift the decimal point *in the direction of* the desired unit. ➞ 0.000020 A or 20 × 10^{-6} A

Decimal point moved six places to the left

FIGURE 1-38 Simple conversion scale for large or small numbers.

Scientific Notation

Using powers of 10 or scientific notation greatly simplifies math operations. A number that has many zeros to the right or to the left of the decimal point is made simpler by putting it in scientific notation. For example, $0.0000035 \times 0.000025$ is difficult to multiply. It can be put in the form $(3.5 \times 10^{-6}) \times (2.5 \times 10^{-5})$. Notice the number of places the decimal point is moved in each number.

Figure 1-39 lists some of the powers of 10. In a whole number, the power to which the number is raised is positive. It equals the number of zeros following the 1. In decimals, the power is negative and equals the number of places the decimal point is moved to the left of the 1. Easy powers of 10 to remember are $10^2 = 100$ (10×10) and $10^3 = 1000$ ($10 \times 10 \times 10$).

Any number written as a multiple of a power of 10 and a number between 1 and 10 is said to be expressed in scientific notation. For example:

$$81,000,000 = 8.1 \times 10,000,000, \text{ or } 8.1 \times 10^7$$

$$500,000,000 = 5 \times 100,000,000, \text{ or } 5 \times 10^8$$

$$0.0000000004 = 4 \times 0.0000000001, \text{ or } 4 \times 10^{-10}$$

Scientific notation simplifies multiplying and dividing large numbers of small decimals. For example:

$$4800 \times 0.000045 \times 800 \times 0.0058$$

$$= (4.8 \times 10^3) \times (4.5 \times 10^{-5}) \times (8 \times 10^2) \times (5.8 \times 10^{-3})$$

$$= (4.8 \times 4.5 \times 8 \times 5.8) \times (10^{3-5+2-3})$$

$$= 1002.24 \times 10^{-3}$$

$$= 1.00224$$

$$95,000 \div 0.0008 = \frac{9.5 \times 10^4}{8 \times 10^4}$$

$$\frac{9.5 \times 10^{4-(-4)}}{8}$$

$$\frac{9.5 \times 10^8}{8} = 1.1875 \times 10^8$$

$$= 118,750,000$$

With some practice the use of scientific notation becomes easy.

	Number	Power of 10
Whole numbers	1,000,000	10^6
	100,000	10^5
	10,000	10^4
	1000	10^3
	100	10^2
	10	10^1
	1.0	10^0
Decimals	0.1	10^{-1}
	0.01	10^{-2}
	0.001	10^{-3}
	0.0001	10^{-4}
	0.00001	10^{-5}
	0.000001	10^{-6}

FIGURE 1-39 Powers of 10.

Unit Conversion

Complete each electric unit problem.

46. 0.65 A = —————— mA.

47. 0.12 μF = —————— pF

48. 0.215 mV = —————— V

49. 0.0000005 F = —————— μF

50. 255 mA = —————— A

51. 45,000 Ω = —————— MΩ

52. 0.85 MΩ = —————— Ω

53. 6500 W = —————— kW

54. 68,000 V = —————— kV

55. 9200 W = —————— MW

Scientific Notation

Write the following numbers as powers of 10.

56. 0.00001 ——————

57. 0.00000001 ——————

58. 10,000,000 ——————

59. 1000 ——————

60. 10 ——————

61. 0.01 ——————

62. 10,000 ——————

63. 0.0001 ——————

64. 1.0 ——————

65. 1,000,000 ——————

Write the following numbers in scientific notation (as the product of a number between 1 and 10 and a power of 10).

66. 0.00128 ——————

67. 1520 ——————

68. 0.000632 ——————

69. 0.0030 ——————

70. 28.2 ——————

71. 7,300,000,000 ——————

72. 52.30 ——————

73. 8,800,000 ———————

74. 0.051 ———————

75. 0.000006 ———————

Answers

46. 650 mA	47. 120,000 pF
48. 0.000215 V	49. 0.5 μF
50. 0.000255 A	51. 0.045 MΩ
52. 850,000 Ω	53. 6.5 kW
54. 68 kV	55. 0.0092 MW
56. 10^{-5}	57. 10^{-8}
58. 10^7	59. 10^3
60. 10^1	61. 10^{-2}
62. 10^4	63. 10^{-4}
64. 10^0	65. 10^6
66. 1.28×10^{-3}	67. 1.52×10^3
68. 6.32×10^{-4}	69. 3.0×10^{-3}
70. 2.82×10^1	71. 7.3×10^9
72. 5.23×10^1	73. 8.8×10^6
74. 5.1×10^{-2}	75. 6.0×10^{-6}

Batteries

Chemical energy is converted into electric energy with chemical cells. When two or more cells are connected in series or parallel or a combination of the two, they form a *battery*. A cell is made of two different metals immersed in a liquid or paste called an *electrolyte*. Chemical cells are either primary or secondary cells. *Primary cells* are usable only for a certain time period. *Secondary cells* are renewed after being used to produce electric energy. This is known as *charging*. Both primary and secondary cells have many uses.

Primary Cells

The operation of a primary cell involves placing two unlike materials called *electrodes* into the *electrolyte*. When the materials of the cell are brought together, their molecular structures change.

(a)

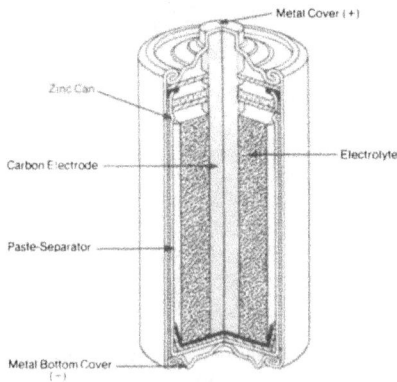

(b)

FIGURE 1-40 Carbon-zinc cell. (a) Pictorial. (b) Cutaway.

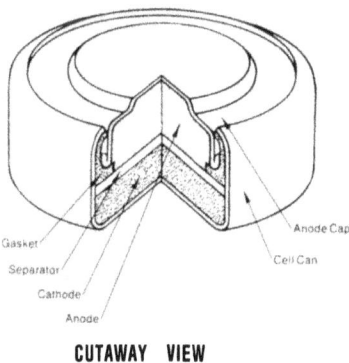

CUTAWAY VIEW

FIGURE 1-41 Mercury cell.

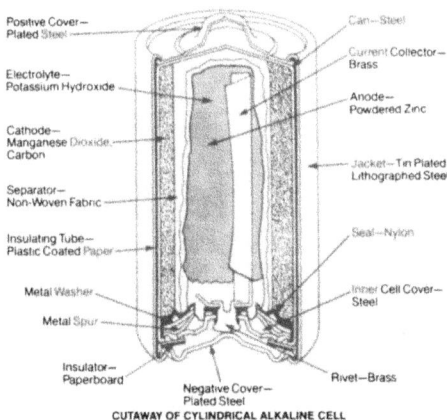

CUTAWAY OF CYLINDRICAL ALKALINE CELL

FIGURE 1-42 Alkaline cell.

During this chemical change, atoms may either gain additional electrons or leave behind some electrons. These atoms then have either a positive or a negative electric charge. They are called *ions.* Ionization of atoms allows the chemical solution of a cell to produce an electric current.

A load device such as a lamp may be connected to a cell. Electrons flow from one of the electrodes of the cell to the other through the electrolyte material. This produces an electric current flow through the load. Current leaves the cell through its negative electrode. It passes through the load device and then goes back to the cell through its positive electrode. A complete circuit exists between the cell (source) and the lamp (load).

The voltage output of a primary cell depends on the electrode materials used and the type of electrolyte. The familiar carbon-zinc cell is shown in Fig. 1-40. It produces approximately 1.5 V. The negative electrode of this cell is the zinc container. The positive electrode is a carbon rod. A paste material acts as the electrolyte. It is placed between the two electrodes. This type of cell is called a *dry cell.*

Many types of primary cells are used today. The carbon-zinc cell is the most common. This cell is inexpensive and is available in many sizes. Applications are mainly for portable equipment and instruments. For uses that require higher voltage or current than one cell can deliver, several cells are combined in series, parallel, or series-parallel connections. Carbon-zinc batteries are available in many voltage ratings.

Another type of primary cell is the mercury (zinc-mercuric oxide) cell shown in Fig. 1-41. This cell is an improvement over the carbon-zinc cell. It has a more constant voltage output, a longer life, and a smaller size. Mercury cells are more expensive than carbon-zinc cells. They produce a voltage of about 1.35 V, which is slightly less than that of carbon-zinc cells.

An alkaline (zinc-manganese dioxide) cell is shown in Fig. 1-42. Alkaline cells have a voltage per cell of 1.5 V. They supply higher-current electric loads. They have much longer lives than carbon-zinc cells of the same types.

Secondary Cells

Chemical cells that may be reactivated by means of charging are called *secondary cells,* or *storage cells.* Common types are the lead-acid, nickel-iron (Edison), and nickel-cadmium cells. The latter two cells serve as primary electric sources for industry.

Lead-Acid Cells

The lead-acid cell of Fig. 1-43 is a secondary cell. The electrodes of lead-acid cells are made of lead and lead peroxide. The positive plate is lead peroxide (PbO_2). The negative plate is lead (Pb).

The electrolyte is sulfuric acid (H_2SO_4). When the lead-acid cell supplies current to a load, the chemical process is written as follows:

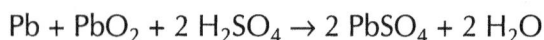

$$Pb + PbO_2 + 2\ H_2SO_4 \rightarrow 2\ PbSO_4 + 2\ H_2O$$

or

Lead plus lead peroxide plus two parts sulfuric acid yields two parts lead sulfate and two parts water

The sulfuric acid ionizes to produce four positive hydrogen ions (H^+) and two negative sulfate (SO_4^{2-}) ions. A negative charge is developed on the lead plate when an SO_4^{2-} ion combines with the lead plate to form lead sulfate ($PbSO_4$). The positive hydrogen ions (H^+) combine with electrons of the lead peroxide plate. They become neutral hydrogen atoms. The H^+ ions also combine with the oxygen (O) of the lead peroxide plate to become water (H_2O). The lead peroxide plate then has a positive charge. A lead-acid cell has a voltage between electrodes of about 2.1 V when fully charged.

Cells discharge when supplying current for long periods of time. They are no longer able to develop an output voltage when discharged. Cells may be charged by causing direct current to flow through the cell in the opposite direction. The chemical process of charging is written as

$$2\ PbSO_4 + 2H_2O \rightarrow Pb + PbO_2 + 2\ H_2SO_4$$

or

Two parts lead sulfate plus two parts water yields lead plus lead peroxide plus two parts sulfuric acid

The original condition of the chemicals is reached through charging. The chemical reaction is reversible.

The amount of charge of a lead-acid cell is measured with a *specific gravity* test. A *hydrometer* is an instrument used to test the electrolyte solution. The specific gravity of a liquid is an index of how heavy a liquid is compared with water. Pure sulfuric acid has a specific gravity of 1.840. The dilute sulfuric acid of a fully charged lead-acid cell varies from 1.275 to 1.300. During discharge of the cell, water is formed. This reduces the specific gravity of the electrolyte. A specific gravity between 1.120 and 1.150 indicates a fully charged cell.

The *capacity* of a battery made of lead-acid cells is given by an *ampere-hour (A-h) rating*. A 50 A-h battery is rated to deliver 50 A for 1 hour, 25 A for 2 hours, or 12.5 A for 4 hours. The ampere-hour rating is an approximate value. It depends on the rate of discharge and the operating temperature of the battery.

Nickel-Cadmium Cells

Another type of secondary cell is a *nickel-cadmium* cell. These

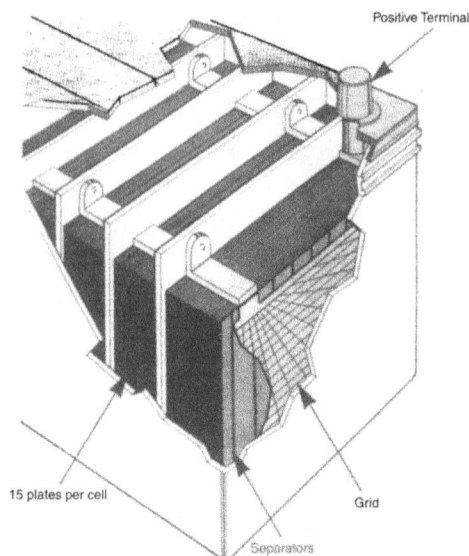

FIGURE 1-43 Lead-acid cells of a battery. (Courtesy of Exide Corp.)

cells are available in many sizes. They are often used in portable equipment. The positive plate of this cell is nickel hydroxide. The negative plate is cadmium hydroxide. The electrolyte is made of potassium hydroxide. These cells have a long life. A fully charged nickel-cadmium cell has a voltage of approximately 1.25 V.

Other Secondary Cells

There are other types of secondary cells used today. Many have specialized applications. Among these are silver oxide–zinc cells and silver-cadmium cells. These cells have a high output and long life. They are more expensive than other types of secondary cells of the same size.

Secondary cells have many uses. Storage batteries are used in some buildings to provide emergency power when a power failure occurs. Standby systems are needed, especially for lighting when power is off. Automobiles use storage batteries for everyday operation. Many types of instruments and portable equipment use batteries for power. Some instruments use rechargeable secondary cells and others use primary cells.

Battery Connections

Each electric circuit requires a voltage source. One source for dc circuits is a cell or battery. The arrangement of the cells in a circuit depends on the *load* requirements of voltage and current. If the voltage must be high, cells are connected in series. Series circuits are discussed in detail in Unit 3.

Series Connection

The voltage of a single primary cell, or dry cell, is 1.5 V. When the voltage required by a load is higher than 1.5 V, it is necessary to use more than one cell, and the cells must be connected in series, as shown in Fig. 1-44. The negative terminal of the first cell is connected to the positive terminal of the second cell. The negative terminal of the second cell is connected to the positive terminal of the third cell, and so on. The positive terminal of the first cell and the negative terminal of the last cell become the output terminals for the circuit. Figure 1-44 is a schematic of four cells in series. The long vertical line represents the positive terminal of each cell, and the short vertical line represents the negative terminal of each cell. When cells are connected in series, the *same* amount of current flows through each cell. The total voltage of the cells connected in series is equal to the *sum* of the voltages of the individual cells.

Parallel Connection

If the current requirement of a circuit is high, cells are connected in parallel. Cells also are rated by current. The current rating of cells is based on a cell's capacity or ability to furnish a certain amount of current for a length of time. The lifetime of cells can be increased by means of connecting more cells in parallel.

Figure 1-45 shows four 1.5 V dry cells connected in parallel. All the positive terminals are connected. Likewise, all the negative terminals are connected. Figure 1-45 shows a schematic of

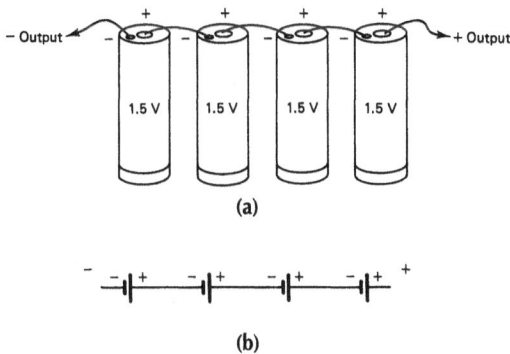

FIGURE 1-44 Series voltage connection. (a) Pictorial. (b) Schematic.

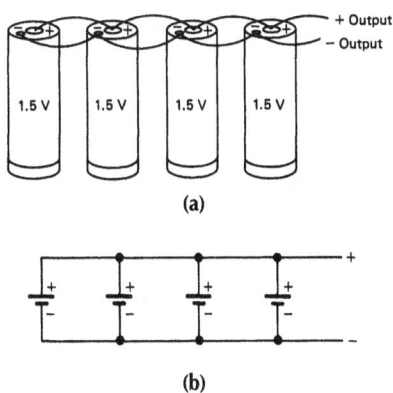

FIGURE 1-45 Parallel voltage connection. (a) Pictorial. (b) Schematic.

the four cells in parallel. When cells are connected in parallel, the total current capacity is equal to the *sum* of the currents of the individual cells. When cells are connected in parallel, the voltage applied to the circuit is the *same* as the voltage of one cell.

Combination (Series-Parallel) Connection

If both the voltage and current requirements of an electric circuit are higher than the rated voltage and current of a single cell, it is necessary to use three or more cells in a series-parallel or combination circuit. Figure 1-46a shows four 1.5 V dry cells connected in series parallel. Two pairs of cells are connected in series, negative to positive. Then the two pairs are connected in parallel, negative to negative and positive to positive. Figure 1-46b is a schematic of the four cells in series parallel. When the cells are connected in series parallel, the voltage applied to the circuit is equal to the *sum* of the cells connected in series. When the cells are connected in series parallel, the total current capacity is equal to the *sum* of the current ratings of the cells connected in parallel.

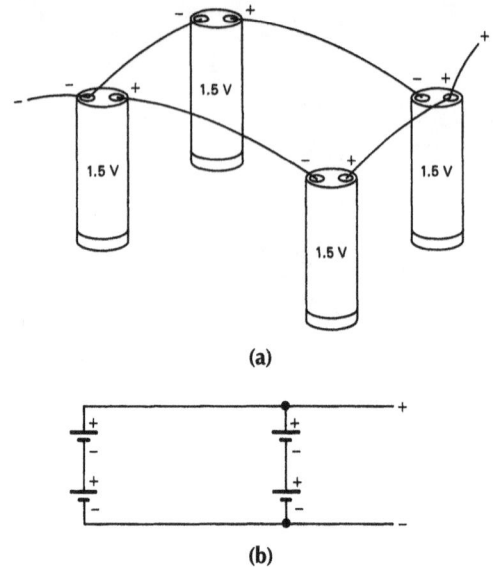

(a)

(b)

FIGURE 1-46 Series parallel voltage connection. (a) Pictorial. (b) Schematic.

Self-Examination

76. A cell that can be recharged is a —————— cell.

77. A low-cost primary cell is called a —————— cell.

78. The total voltage of four 1.5 V cells connected in series is —————— V.

79. The total voltage of two 6 V batteries connected in series is —————— V.

80. When more than 1.5 V is required, cells are connected in ——————.

81. A —————— cell cannot be recharged.

82. A primary cell is destroyed in use and cannot be ——————.

Answers

76. Secondary	77. Carbon-zinc
78. 6	79. 12
80. Series	81. Primary
82. Recharged	

Experimental Activities for DC Electronics

The experimental activities that follow emphasize the practical applications of electronics. They parallel the content of each of the units in this book. The expense of the equipment is kept to a minimum. A few of the activities require no lab equipment. Each experimental activity is organized in the following way:

EXPERIMENT 1-1

(First activity of Unit 1)

TITLE

(Topic of the activity)

Introductory paragraphs containing an overview of the activity, practical applications, the purpose of the activity, and suggested observations that should be made.

OBJECTIVE

Expected learning to take place when the experiment is completed.

EQUIPMENT

Necessary equipment and materials to perform the experiment.

PROCEDURE

Logical step-by-step sequence for completing the learning activity. Maximum use is made of charts and tables that will aid in the recording of data.

ANALYSIS

Specific questions and problems that supplement the experimental activity.

The experimental material is presented with a single-concept approach. Activities organized this way require only a short time to assemble and make the necessary measurements to facilitate learning.

In this book several experimental activities are used to reinforce the text material. These activities provide a different direction for the learning process. As a rule, the activity is experimentally based. This involves some manipulative activity or hands-on operation. The experiments deal with activities such as circuit construction, testing operations, calculations, instrument use, and component identification and use. Through this approach you will become more familiar with electronic components and their use in a specific circuit application.

Tools and Equipment

A variety of tools and components are needed to perform the experimental activities of this course. These may be obtained from electronics supply houses, mail-order supply houses, and educational vendors. A listing of these sources appears in appendix C. These components may be obtained through a variety of electronics supply houses. As a rule, a standard part number must be used in obtaining these components. In many cases, an equivalent component may be made by another manufacturer. Most of the experiments in this book can be completed with a 6 V battery, 1.5 V dry cells, and some other very inexpensive components.

Important Information

At this time you may want to turn to the back of the book and review the following information.

Appendix A: Electronics Symbols

Appendix D: Soldering Techniques

The information in these sections will help you to perform the experimental activities in this book.

Lab Activity Troubleshooting and Testing

The lab activities included in this book provide an opportunity to practice troubleshooting and testing for electronic circuits, devices, and systems. This section of the book provides a comprehensive list of troubleshooting and testing procedures that may be accomplished while performing the lab activities. Emphasis is placed on understanding circuit operation, safety, and proper use of test equipment. A technician who understands how the circuit, device, or system functions and knows how to use test equipment will find troubleshooting and testing relatively easy. This is true for the simplest type of electronic circuit and for complex systems.

Competencies for Troubleshooting and Testing

Specific competencies developed through the use of this book are listed here. These competencies are achieved through troubleshooting and testing procedures. The laboratory activities provide a way to learn these procedures. The competency list identifies specific procedures that are being mastered. Competencies are divided according to the lab activity in which they are introduced. Some troubleshooting procedures may be repeated several times. The competency list is highly important in the study of troubleshooting and testing. You should mark this list of activities and refer to it as you progress through the book.

OBJECTIVES

Upon completing the activities presented in this book, you should be able to do the following:

1. Outline basic troubleshooting procedures for locating specific trouble with devices and equipment.

2. Find the parts or circuits that are defective by using a common-sense approach.

3. Test devices and circuits of electronic equipment using correct procedures.

Competency List

Unit 1: Direct Current (DC) Electronics

EXPERIMENT 1-1—COMPONENTS, EQUIPMENT, AND SYMBOLS

1. Recognize electronics symbols used with schematics.

2. Recognize electronic devices that are commonly used in circuits.

3. Construct a simple electronic circuit by using a schematic.

EXPERIMENT 1-2—RESISTOR COLOR CODE

1. Recognize value of resistors that have a color code.

Unit 2: Measuring Voltage, Current, and Resistance

EXPERIMENT 2-1—MEASURING RESISTANCE

1. Measure resistance with a multimeter.

2. Test the resistance value of a fixed resistor.

3. Test the resistance value and operating condition of a potentiometer.

4. Learn to read the resistance scale and select the proper range of an analog multimeter.

5. Test for open circuits and short circuits (continuity).

EXPERIMENT 2-2—MEASURING VOLTAGE

1. Measure dc voltage with a multimeter.

2. Test the voltage of a battery (dc source).

3. Test for open circuits with a voltmeter.

4. Test for short circuits with a voltmeter.

5. Learn to read the dc voltage scale and select the proper range.

6. Learn to observe polarity when taking measurements of dc circuits.

EXPERIMENT 2-3—MEASURING CURRENT

1. Measure dc current with a multimeter.

2. Test the amount of total current flowing from a dc source.

3. Test for open circuits with an ammeter.

4. Test for short circuits with an ammeter.

5. Learn to read the dc amps (or milliamps) scale and select the proper range of an analog multimeter.

6. Learn to observe polarity when taking dc current measurements.

EXPERIMENT 2-4—FAMILIARIZATION WITH POWER SUPPLY

1. Operate a variable dc power supply.

2. Use a battery and potentiometer circuit as a variable dc power supply.

Unit 3: Ohm's Law and Electric Circuits

EXPERIMENT 3-1—APPLICATION OF OHM'S LAW

1. Apply Ohm's law to solve dc circuit problems.

2. Compare calculated and measured values of voltage and current.

3. Learn the effect of increasing or decreasing voltage, current, or resistance on the operation of a dc circuit.

4. Measure voltage, current, and resistance of a dc circuit with a multimeter.

EXPERIMENT 3-2—SERIES DC CIRCUITS

1. Construct a series dc circuit.

2. Measure voltage, current, and resistance of a series dc circuit with a multimeter.

3. Compare measured and calculated values of voltage, current, and resistance of a series dc circuit.

4 Recognize the characteristics of a series dc circuit in terms of voltage drop, current flow, and total resistance.

EXPERIMENT 3-3—PARALLEL DC CIRCUITS

1. Construct a parallel dc circuit.

2. Measure voltage, current, and resistance of a parallel dc circuit with a multimeter.

3. Compare measured and calculated values of voltage, current, and resistance of a parallel dc circuit.

4. Recognize the characteristics of a parallel dc circuit in terms of voltage drop, current flow, and total resistance.

EXPERIMENT 3-4—COMBINATION DC CIRCUITS

1. Construct a combination dc circuit.

2. Measure voltage, current, and resistance of a combination dc circuit with a multimeter.

3. Compare measured and calculated values of voltage, current, and resistance of a combination dc circuit.

4. Recognize the characteristics of a combination dc circuit in terms of voltage drop, current flow, and total resistance.

EXPERIMENT 3-5—POWER IN DC CIRCUITS

1. Calculate power values in a dc circuit.

2. Use a multimeter to make voltage and current measurements to verify power calculations.

3. Determine maximum current and power rating of electronic components.

EXPERIMENT 3-6—VOLTAGE DIVIDER CIRCUITS

1. Study the characteristics of a potentiometer in terms of resistance.

2. Use a potentiometer as a voltage divider.

3. Calculate resistance values to design a voltage divider.

4. Use a multimeter to make voltage measurements on a voltage divider circuit.

EXPERIMENT 3-7—KIRCHHOFF'S VOLTAGE LAW

1. Apply Kirchhoff's voltage law for circuits with one or two voltage sources.

2. Use a multimeter to measure voltage values for applying Kirchhoff's voltage law.

3. Compare measured and calculated values of voltage and current after applying Kirchhoff's voltage law.

EXPERIMENT 3-8—KIRCHHOFF'S CURRENT LAW

1. Apply Kirchhoff's current law for dc circuits.

2. Use a multimeter to measure current values for applying Kirchhoff's current law.

3. Compare measured and calculated values of current after applying Kirchhoff's current law.

EXPERIMENT 3-9—SUPERPOSITION METHOD

1. Apply the superposition method to solve dc circuit problems.

2. Use a multimeter to measure voltage and current values for applying the superposition method.

3. Compare measured and calculated values of current after applying the superposition method.

EXPERIMENT 3-10—THEVININ EQUIVALENT CIRCUITS

1. Apply Thevinin's method to solve for equivalent voltage and resistance values in one- and two-source dc circuits.

2. Use a multimeter to measure voltage and resistance values for applying Thevinin's method.

3. Compare measured and calculated values of equivalent voltage and resistance in one- and two-source dc circuits.

EXPERIMENT 3-11—NORTON EQUIVALENT CIRCUITS

1. Apply Norton's method to solve for equivalent current and resistance values in a dc circuit.

2. Use a multimeter to measure current and resistance values for applying Norton's method.

3. Compare measured and calculated values of equivalent current and resistance of a dc circuit.

EXPERIMENT 3-12—MAXIMUM POWER TRANSFER

1. Apply the maximum power transfer principle to a dc circuit.

2. Learn that maximum power will transfer from the source to the load of a circuit when source resistance equals load resistance.

3. Use a multimeter to measure voltage and resistance values for studying maximum power transfer.

4. Calculate power values as load resistance changes in value.

5. Plot a graph that shows the relation of load resistance value to power output of a dc circuit.

EXPERIMENT 3-13—BRIDGE CIRCUITS

1. Apply bridge circuit simplification procedures to a dc circuit.

2. Use a multimeter to measure voltage, current, and resistance values of a bridge circuit.

3. Compare measured and calculated values of voltage, current, and resistance in bridge circuits.

Unit 4: Magnetism and Electromagnetism

EXPERIMENT 4-1—THE NATURE OF MAGNETISM

1. Determine the polarity of a magnet with a compass.

2. Observe the reaction of a magnetic field for attraction and repulsion.

EXPERIMENT 4-2—ELECTROMAGNETIC RELAYS

1. Measure the resistance of a relay coil with a multimeter.

2. Identify the contacts of a relay.

3. Construct a relay circuit.

4. Measure the pickup and dropout current of a relay.

Unit 5: Electronic Instruments

No experiments. Instruments are integrated into other activities.

Unit 6: Inductance and Capacitance

EXPERIMENT 6-1—TIME-CONSTANT CIRCUITS

1. Construct an *RC* circuit and compute the charge and discharge time.

2. Measure the charge and discharge current of an *RC* circuit with a multimeter.

3. Measure the voltage and current changes that occur for different time constants.

EXPERIMENT 1-1

COMPONENTS, EQUIPMENT, AND SYMBOLS

Basic to the study of any subject is understanding of the language and symbols used. The study of dc electronics has its own language, both graphic and verbal. The graphic language used with dc electronics includes symbols that represent components, devices, and equipment.

OBJECTIVES

1. To examine and become familiar with some of the graphic symbols used in dc electronics.

2. To become familiar with and learn to recognize symbols used for common devices.

EQUIPMENT

Multimeter (VOM)

SPST switch

Potentiometer (any value)

Connecting wires for circuit board

6 V battery

6 V lamp with socket

PROCEDURE

1. The connecting wires you will use in this activity are paths for the movement of electrons. These connecting wires may be connected to each other at almost any angle or may cross each other without being connected. The graphic symbols representing these connecting wires, called *conductors*, are illustrated in Fig. 1-1A.

2. In Fig. 1-1B, how many times do conductors cross each other and how many times are conductors connected to each other? Conductors cross each other _____ times. Conductors are connected to each other _____ times.

3. A single-pole, single-throw (SPST) switch is a device used to allow current to flow when closed, or on, and interrupt the flow of current when open, or off. A switch is used to turn on and turn off the light in a room. The symbols representing the SPST switch in its two conditions, on and off, are illustrated in Fig. 1-1C.

4. Fixed resistors are some of the most widely used components to control current flow. The symbol for a fixed resistor is illustrated in Fig. 1-1D.

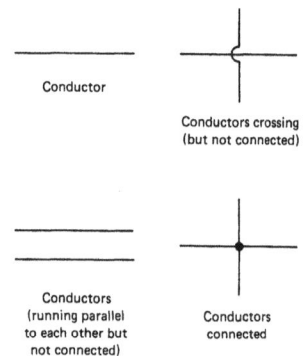

Conductor

Conductors crossing (but not connected)

Conductors (running parallel to each other but not connected)

Conductors connected

FIGURE 1-1A

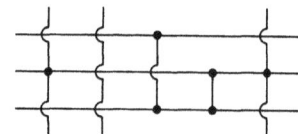

FIGURE 1-1B

SPST "on" or "closed"

SPST "off" or "open"

FIGURE 1-1C

FIGURE 1-1D

FIGURE 1-1E

Potentiometer   Potentiometer

FIGURE 1-1F

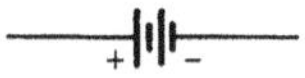

+ ╢╟ −
Battery

FIGURE 1-1G

5. A light bulb or lamp is a common component widely used in flashlights and other devices to produce light. The symbol for an incandescent lamp is shown in Fig. 1-1E.

6. A potentiometer, or pot, is a resistive component that can be adjusted to control current flow. A pot is used to increase or decrease the volume of a radio or TV set. Fig. 1-1F shows the graphic symbols for a potentiometer. The center connection of the pot is designated by an arrow and represents the adjustable portion of the device.

7. A 6 V battery is a chemical source of electric energy that causes current to flow through conductors, resistors, and other components. The symbol in Fig. 1-1G is used for a battery. One of the connectors of the battery is labeled with a + sign and the other is labeled with a − sign. It is always important to determine how the battery is connected to components in a circuit.

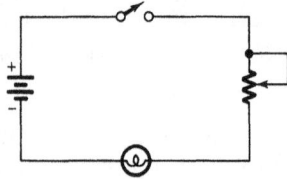

FIGURE 1-1H

8. Use your circuit board or trainer unit to set up the circuit shown in Fig. 1-1H. Some of the types of circuit boards and trainers that may be used for your experiments are shown in Fig. 1-1I. You should study the layout of the one you are using to be sure that you connect this circuit and the others you construct properly. You should become very familiar with the use of a circuit board. For most circuit boards it is necessary to cut wire of the proper diameter to various sizes and to strip about ¼ inch of insulation from both ends. The wires are then used to make circuit connections. The switch and potentiometer should have wires soldered to their terminals about 2 inches long for connecting to the circuit board. They will be used for many other experiments. This would be a good time to review the appendix material on soldering and electronics tools in appendix D and appendix E.

9. Complete the circuit connection using your circuit board. Close the SPST switch and adjust the potentiometer from its maximum counterclockwise position to its maximum clockwise position. What happens to the lamp? If the circuit is constructed properly, the lamp should get brighter and then dimmer as the pot is adjusted.

10. Adjust the potentiometer until the lamp is at its brightest and open the SPST switch. What happens to the lamp?

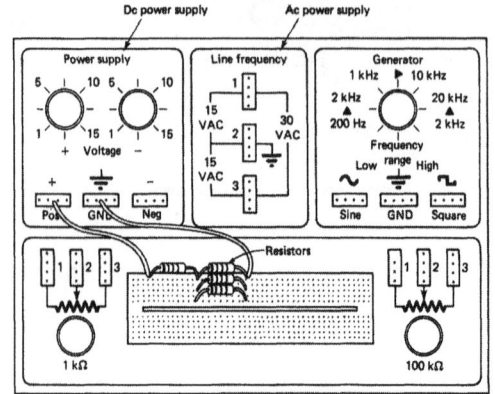

(a)

Quick test sockets and bus strips (various sizes)

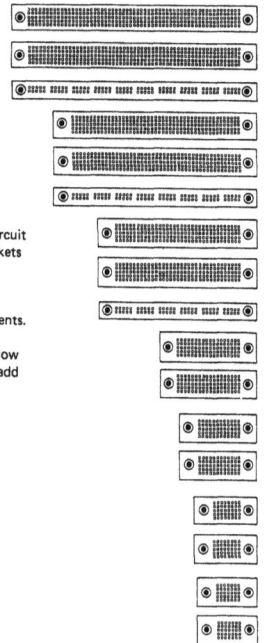

Breadboard any circuit by using these sockets and bus strips. No soldering or patch cords, just plug-in wires and components. These strips snap together, which allow you to remove or add strips for your requirements.

Wire jumper kit*

Use with Quick test sockets and bus strips. Pre-cut, pre-stripped, pre-formed AWG #22, insulated solid hookup wire in color-coded lengths.

*You can make your own!

(b)

FIGURE 1-1I (a) Heath circuit trainer. (b) Individual circuit boards and wires

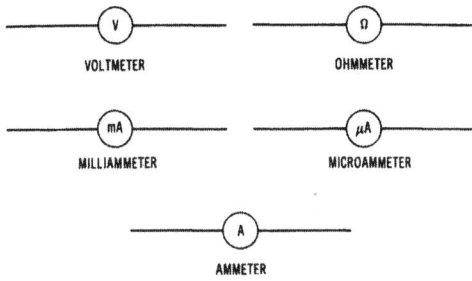

FIGURE 1-1J

11. A meter called a VOM, or multimeter, is designed to measure several electric quantities. Generally a switch called the *function switch* is used to control what the meter is to measure. Most VOMs are capable of measuring volts, ohms, milliamperes, and amperes. The symbols for a VOM as it is adjusted to each function are shown in Fig. 1-1J.

12. For the meter you are using, list the electric quantities your meter will measure. _____

ANALYSIS

1. Draw the symbols for the indicated components in the space provided.

   Fixed resistor:

   Potentiometer:

   Open SPST switch:

2. Draw the symbols for conductors in the space provided.

   Conductors crossing:

   Conductors connected:

3. Draw the symbol for a battery placing the positive (plus) sign and the negative (minus) sign at the proper sides of the symbol.

4. Identify each of the meter symbols illustrated in Fig. 1-1K.

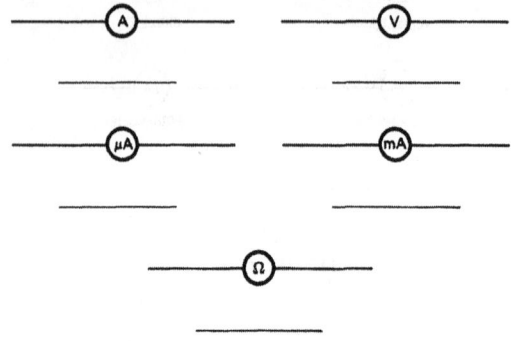

FIGURE 1-1K

RESISTOR COLOR CODE

Many of the components used in electronics have various colors in the form of dots, circles, and bands that indicate values. This arrangement is most commonly used with fixed resistors and capacitors. Simply to print or letter the value of a component on its body is extremely impractical because of the small size of many of the components, the location of components in circuits, or the long numbers that indicate the values of some components.

OBJECTIVE

To examine the color coding system used with fixed resistors.

EQUIPMENT

Fixed resistors (eight of any value)

PROCEDURE

1. The most important things to learn when dealing with color-coded components are the meaning and placement of the colors on the body of the component. Figure 1-2A shows a fixed resistor that has the band system of color coding.

2. Figure 1-2A illustrates the standard for color-coding fixed resistors. The organizations that developed these standards are the military and the Electronic Industries Association (EIA). Both standards use the same 12 basic colors to designate the value of a fixed resistor. In both systems the *first color* indicates the first digit in the value of the fixed resistor, which is measured in ohms. The *second color* indicates the second digit in the value of the resistor. The *third color* indicates the number by which the first two digits are to be multiplied, or the number of zeros to be added to the first two numbers. The *fourth color*, if there is a fourth color, indicates the *tolerance* of the value of the resistor. Tolerance is expressed as ±5%, ±10%, or ±20% of the value in ohms indicated by the first three colors. A *fifth color*, which is seldom used, indicates the failure rate per 1000 hours of use associated with the resistor and is discussed herein.

3. Figure 1-2B shows a resistor that has the band-color-code marking system. Read the colors from left to right, or from the end of the resistor closest to where the bands begin. Using Fig. 1-27 you should be able to determine the value of the resistor in ohms and its tolerance.

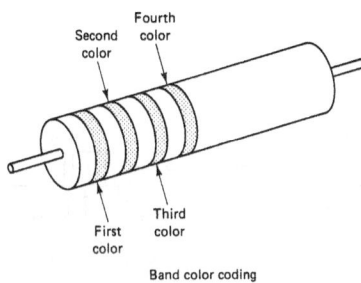

Band color coding

FIGURE 1-2A

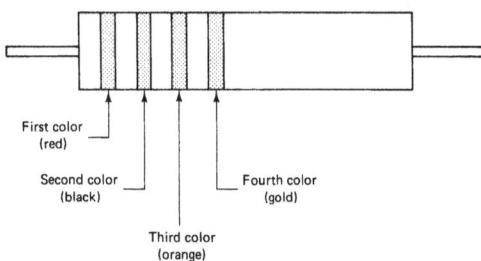

First color (red)

Second color (black)

Third color (orange)

Fourth color (gold)

FIGURE 1-2B

The first color is *red,* which indicates the first digit in the value of the resistor is 2.

The second color is *black,* which indicates the second digit in the value of the resistor is 0. We now have the first two digits of the ohmic value of the resistor, which is 20.

The third color is *orange,* which indicates the number by which the first two digits are to be multiplied or the number of zeros to add to the first two digits. Because orange designates a multiplier of 1000, we can multiply 20 by 1000 to give us the value in ohms of the resistor ($20 \times 1000 = 20,000 \; \Omega$). Orange also indicates that we add three zeros to the first two digits to determine the ohmic value of the resistor (20 with three zeroes added gives 20,000 Ω). The two methods are equally effective.

The fourth color is *gold,* which indicates the tolerance of the resistor. The tolerance shows how close the measured value of the resistor is to the rated value and is normally expressed as a percentage of the rated value. In this case, gold indicates a tolerance of ±5%. This means this resistor will be within 5% of 20,000 Ω one way or the other. Thus the value of the illustrated resistor can be as low as 19,000 ohms ($20,000 \times 0.05 = 1000$; $20,000 - 1000 = 19,000$) or as high as 21,000 ($20,000 \times 0.05 = 1000$; $20,000 + 1000 = 21,000$).

4. In some instances, a fourth color is not indicated on the body of the resistor. When this condition occurs, you may automatically assume its tolerance to be 20%. Frequently no fifth color appears on the resistor. There is no accurate method of predicting failure rate without the manufacturer's specifications. In Fig. 1-2C, indicate the values of the illustrated resistors with ohmic value and tolerance.

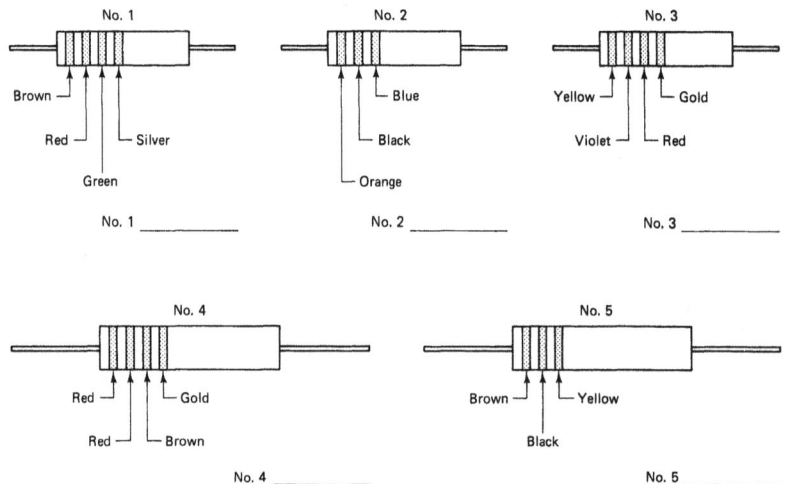

FIGURE 1-2C

5. You should now be able to determine the value and tolerance of any color-banded resistor by using Fig. 1-27. Without the table, you might be helpless. Another way to determine the basic value of any color-coded resistor is to remember the following quotation: "Big brown rabbits often yield great big vocal groans when gingerly slapped." The first letter of each word in this statement corresponds to the first letter of each of the 12 colors used in the color code. In addition, the words of the statement may be counted, beginning with zero to determine which word corresponds with each digit or the number of zeros to be added to complete the value of the resistor. Using the quotation, determine the values of eight resistors in your component package.

6. In the following spaces, list, in order, the color bands of each of the eight resistors. After listing the colors, in order, record the value and tolerance of each resistor. Example:

Example: Brown, Black, Gold, Red

   10 Ω,   5%

a. ———, ———, ———, ———

   __Ω, __%

b. ———, ———, ———, ———

   __Ω, __%

c. ———, ———, ———, ———

   __Ω, __%

d. ———, ———, ———, ———

   __Ω, __%

e. ———, ———, ———, ———

   __Ω, __%

f. ———, ———, ———, ———

   __Ω, __%

g. ———, ———, ———, ———

   __Ω, __%

h. ———, ———, ———, ———

   __Ω, __%

ANALYSIS

1. What are the two organizations that developed the systems of resistor color coding? _____

2. What is represented by the first and second colors of a color-banded resistor? _____

3. What is indicated by the third color of a color-banded resistor? _____

4. What unit of electric measurement is used to indicate resistance? _____

5. What is indicated by the fourth color of a color-banded resistor? _____

6. If no fourth color is indicated on the body of a resistor, what is its tolerance value? _____

Unit 1 Examination

Basics of DC Electronics

Instructions: For each of the following, circle the answer that most correctly completes the statement.

1. One coulomb per second is equal to one
   a. Volt
   b. Ohm
   c. Ampere
   d. Farad

2. The force that causes current to flow through a resistance is measured in
   a. Ohms
   b. Amperes
   c. Volts
   d. Henrys

3. The center of an atom is known as the
   a. Nucleus
   b. Neutron
   c. Electron
   d. Proton

4. A circuit that develops 5 mA of current flow has how many amperes of current flow?
   a. 0.5 A
   b. 0.05 A
   c. 0.005 A
   d. 0.0005 A

5. A circuit that develops 1000 µA of current flow is said to have how many amperes of current flow?
   a. 1 A
   b. 0.01 A
   c. 0.001 A
   d. 0.0001 A

6. The total number of protons and neutrons of an atom is known as its
   a. Atomic number
   b. Mass number
   c. Ionic number
   d. Isotopic number

7. The maximum number of electrons in shell K of an atom would be
   a. 2
   b. 4
   c. 6
   d. 8

8. Static electricity is often produced by
   a. Pressure
   b. Magnetism
   c. Heat
   d. Friction

9. Resistance is measured in
   a. Coulombs
   b. Henrys
   c. Ohms
   d. Watts

10. The nucleus of an atom contains
    a. Electrons and neutrons
    b. Protons and neutrons
    c. Protons and electrons
    d. Protons, electrons, and neutrons

11. The best conductor of electric current is
    a. Zinc
    b. Copper
    c. Aluminum
    d. Silver

12. Anything that occupies space and has weight is known as
    a. A substance
    b. A compound
    c. An atom
    d. Matter

13. The term *current* refers to the
    a. Flow of electrons along a conductor
    b. Voltage applied to a circuit
    c. Opposition to flow of electrons
    d. Electron movement from positive to negative

14. A variable resistor is known as
    a. A fixed resistor
    b. An insulator
    c. A capacitor
    d. A potentiometer

15. The plates of a storage battery are made of
    a. Nickel and iron oxide
    b. Cadmium and chromium
    c. Steel and powdered iron
    d. Lead and lead peroxide

16. Two or more cells connected in series compose
    a. A battery
    b. A solar cell
    c. An energizer
    d. An alkaline cell

17. A resistor color coded brown, black, brown has a value of
    a. 100 Ω
    b. 1000 Ω
    c. 10 Ω
    d. 1 Ω

18. A gold tolerance band on a resistor indicates a tolerance of
    a. 1%
    b. 5%
    c. 10%
    d. 20%

19. A resistor color coded black, brown, black has a value of
    a. 0 Ω
    b. 1 Ω
    c. 10 Ω
    d. 100 Ω

20. A cell that may be recharged is classified as
    a. A primary cell
    b. A secondary cell
    c. An energizer cell
    d. An alkaline cell

True or False: Place either T or F in each blank.

_____ 21. A complete electric system contains a source, conductor, control, and load.

_____ 22. Matter exists only in solid form.

_____ 23. An insulator is a material through which an electric current may flow easily.

_____ 24. The term voltage refers to a difference in potential between two points in an electric circuit.

_____ 25. Unlike electric charges repel each other.

Measuring Voltage, Current, and Resistance

Another important activity in the study of electronics is measurement. Measurements are made in many types of electronic circuits. The proper ways of measuring resistance, voltage, and current should be learned. These are the three most common electric measurements.

UNIT OBJECTIVES

Upon completing this unit, you will be able to do the following:

1. Connect an ammeter in a circuit and measure current.

2. Demonstrate how the voltmeter, ammeter, and ohmmeter are connected to a circuit.

3. Measure current, voltage, and resistance of basic electronic circuits.

4. Compare calculated and measured values of a circuit.

5. Demonstrate safety while making electric measurements.

6. Demonstrate proper, safe use of an ohmmeter to measure resistance.

Important Terms

Before reading this unit, review the following terms for a basic understanding of terms associated with electronic measurement.

Ammeter A meter used to measure current (amperes).

Continuity check A test to see whether a circuit is an open or closed path.

Multimeter A meter used to measure two or more electric quantities, such as a volt-ohm-milliammeter (VOM), which measures voltage, resistance, and current, or a digital voltmeter (DVM).

Multirange meter A meter that has two or more ranges to measure an electric quantity.

Ohmmeter A meter used to measure resistance (ohms).

Polarity The direction of an electric potential (– or +) or a magnetic charge (north or south).

Schematic A diagram used to show how the components of electric circuits are wired together.

Voltmeter A meter used to measure voltage.

Volt-ohm-milliammeter (VOM) A multifunction, multirange meter that usually is designed to measure voltage, current, and resistance; also called a multimeter.

Measuring Resistance

Many important electric tests may be made by means of measuring resistance. Resistance is opposition to the flow of current in an electric circuit. The current that flows in a circuit depends on the amount of resistance in that circuit. You should learn to measure resistance in an electric circuit by using a meter.

A volt-ohm-milliammeter (VOM) or multimeter such as the one shown in Fig. 2-1 is often used for doing electric work. A multimeter is used to measure resistance, voltage, or current. The operator changes the type of measurement by adjusting the *function-select switch* to the desired measurement. Figure 2-2 shows the controls of a common type of VOM. This type of meter also is called an *analog meter*. We will discuss the analog meter so you will learn to interpret scales for resistance, voltage, or current measurement. A digital meter uses the same basic rules but is easier to read.

The function-select switch is in the center of the meter. Some of the ranges are for measuring ohms, or resistance. This is called a multirange, multifunction meter or *multimeter*. The ohms measurement ranges are divided into four portions: ×1, ×10, ×1000, and ×100,000. Most VOMs are similar to the example shown.

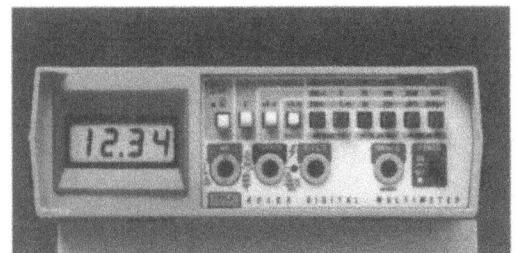

FIGURE 2-1 Meter used to measure electric quantities.

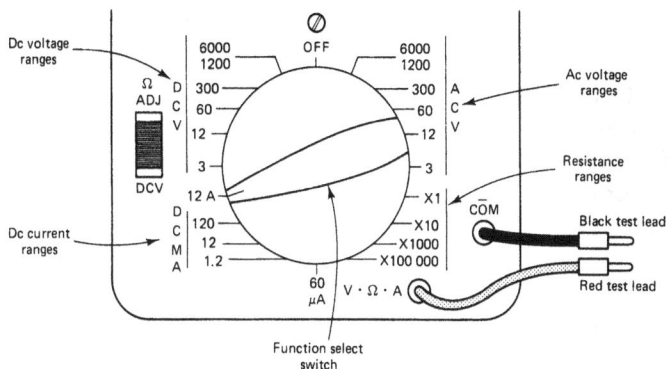

FIGURE 2-2 Controls of a typical VOM.

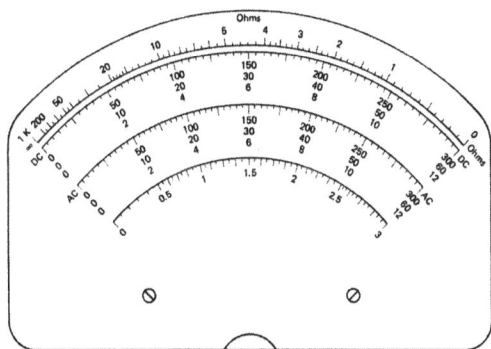

FIGURE 2-3 VOM scale.

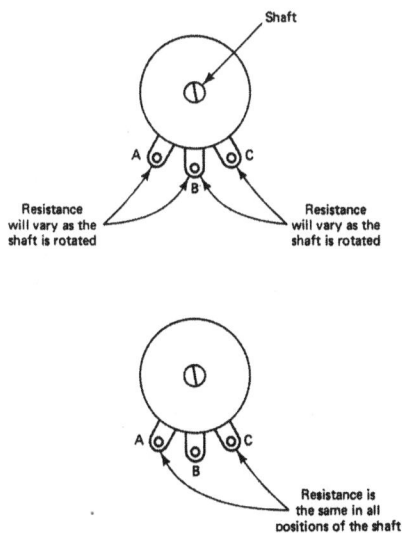

FIGURE 2-4 Measuring the resistance of a potentiometer.

The VOM is adjusted to any of the four positions for measuring resistance. The test leads used with a VOM are ordinarily black and red. These colors are used to help identify which lead is the positive and which is the negative side of the meter. This is important for measuring dc values. Red indicates positive polarity (+) and black indicates negative (–) polarity.

Refer again to Fig. 2-2. The red test lead is put in the hole, or jack, marked with V-Ω-A, or volts-ohms-amperes. The black test lead is put in the hole or jack labeled –COM, or negative common. The function-select switch should be placed on one of the resistance ranges. When the test leads are touched together, or "shorted," the meter needle moves from the left side of the meter scale to the right side. This test shows that the meter is operational.

Now study the scale of the meter. Figure 2-3 shows the scale of one type of VOM. The top scale, from 0 to infinity (∞), is labeled *Ohms*. This scale is used for measuring ohms only. On most VOMs, the top scale is the resistance or ohms scale. To measure any resistance, first select the proper meter range. On the meter range shown in Fig. 2-2 the four ranges, ×1, ×10, ×1000, and ×100,000, are called *multipliers*. The ohmmeter must be properly zeroed before an attempt is made to measure resistance accurately. To *zero* the ohmmeter properly, touch the two test leads together. This should cause the needle to move from infinity (∞) on the left to zero (0) on the right. Infinity represents a very high resistance. Zero represents a very low resistance. If the needle does not reach zero or goes past zero when the test leads are touched or shorted, the control marked *ohms adjust* is used. The needle is adjusted to zero when the test leads are touched together. The ohms-adjust control is indicated by ΩADJ in Fig. 2-2. The ohmmeter should be zeroed before every resistance measurement and after changing ranges. If the meter is not zeroed, measurement will be incorrect.

A more accurate measurement of resistance is made when the meter needle stops somewhere between the center of the ohms scale and zero. Choosing the proper range adjustments controls how far the needle moves. If the range selected is ×1, the number to which the needle points is multiplied by 1. If the function-select switch is adjusted to the ×100,000 range, the number to which the needle points is multiplied by 100,000. Always zero the meter when changing ranges, and always multiply the number indicated on the scale by the multiplier of the range. Do not measure the resistance of a component until it has been disconnected, or the reading may be wrong. Voltage should never be applied to a component when resistance is being measured.

A VOM may be used to measure the resistance of a potentiometer, as shown in Fig. 2-4. If the shaft of the pot is adjusted while the ohmmeter is connected to points *A* and *C*, no resistance change takes place. The resistance of the potentiometer is measured in this way. Connecting to points *B* and *C* or to points *B* and *A* allows changes in resistance as the shaft is turned. The potentiometer shaft may be adjusted both clockwise and counterclockwise. This adjustment affects the measured resistance across points *B* and *C* or *B* and *A*. The resistance varies from zero to maximum and from maximum back to zero as the shaft is adjusted.

How to Measure Resistance with a VOM

Remember that resistance is opposition to the flow of electric current. For example, a lamp connected to a battery has resistance. Its resistance value is determined by the size of the filament wire. The filament wire opposes the flow of electric current from the battery. The battery causes current to flow through the lamp's filament. The amount of current through the lamp depends on the filament resistance. If the filament offers little opposition to current flow from the battery, a large current flows in the circuit. If the lamp filament has high resistance, it offers a great deal of opposition to current flow from the battery. Then a small current flows in the circuit.

Resistance tests are sometimes called *continuity checks*. A continuity check is made to see whether a circuit is open or closed (a continuous path). An ohmmeter also is used to measure exact values of resistance. Resistance always must be measured with no voltage applied to the component being measured. The ohmmeter ranges of a VOM are used to measure resistance. Electrical technicians often use this type of meter because it measures resistance, voltage, or current. When the rotary function-select switch is adjusted, the meter can be set to measure resistance, voltage, or current. The meter switch shown in Fig. 2-2 has the following settings:

1. Direct current (dc) voltage

2. Direct current (dc) amps and milliamps

3. Alternating current (ac) voltage

4. Resistance (ohms)

The lower right part of the function-select switch is for measuring resistance or ohms. The ohms measurement settings are marked as ×1, ×10, ×1000, and ×100,000. When measuring resistance with an ohmmeter, first put the test leads into the meter. The test leads usually are black and red wires that plug into the meter. The red wire is plugged into the hole marked volts-ohms-amps (V-Ω-A). The black wire is plugged into the hole marked negative common (–COM).

The scale of the meter is used to indicate the value of resistance in ohms. The right side of the scale is marked with a zero and the left side is marked with an infinity (∞) sign.

When the test leads are touched together, the needle on the scale of the meter should move to the right side of the scale. This indicates zero resistance. The needle of the meter is adjusted so that it is exactly over the zero mark. This is called *zeroing* the meter. This must be done to measure any resistance accurately. The ohms-adjust (ΩADJ) control is used to zero the needle of the meter. The meter should be zeroed before each resistance measurement is made.

It is important to be able to read the scale of the meter. The top scale shown in Fig. 2-3 is labeled with a zero on the right side and an infinity (∞) sign on the left side. This scale is used for measuring resistance only. The most accurate readings are made when the meter needle moves to somewhere between the center

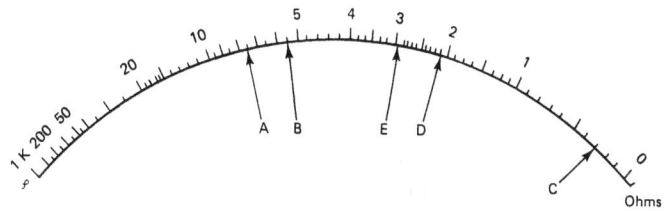

FIGURE 2-5 Ohms scale of a VOM

of the scale and zero because of the greater distance between the numbers on the right of the scale.

To measure any resistance accurately, first select the proper range. The ranges for measuring resistance are on the lower right part of the function-select switch. This switch has resistance ranges marked as ×1, ×10, ×1000, and ×100,000. If the range setting is on ×1, the reading on the meter scale must be multiplied by 1.

The meter must be zeroed whenever a range is changed. The test leads are then placed across a resistance. Assume that the needle of the meter moves to point A on the scale of Fig. 2-5. The resistance equals $7.5 \times 1 = 7.5\ \Omega$. Now change the meter range to ×1000. The reading at point B equals $5.5 \times 1000 = 5500\ \Omega$. At point C, the reading is $0.3 \times 1000 = 300\ \Omega$. The same procedure is used for the ×100,000 range. If the needle moves to 2.2 (point D) on the scale, the reading equals $2.2 \times 100,000$, or $220,000\ \Omega$. If the meter range is set on ×100,000 and the needle moves to 3.9 (point E) on the scale, the reading is $3.9 \times 100,000$, or $390,000\ \Omega$.

Remember to zero the meter by touching the test leads together and use the ohms-adjust control before making a resistance measurement. Each time the meter range is changed, the meter needle must be zeroed on the scale. If this procedure is not followed, the meter reading will not be accurate.

To learn to measure resistance, it is easy to use color-coded resistors. These resistors are small and easy to handle. Practice using the meter to measure several values of resistors makes reading the meter much easier.

Self-Examination

Refer to the Figure below and fill in the blanks with the values that correspond to the pointer location.

Ohms range	A	B	C	D	E
R × 1	(1) ____	(6) ____	(11) ____	(16) ____	(21) ____
R × 10	(2) ____	(7) ____	(12) ____	(17) ____	(22) ____
R × 100	(3) ____	(8) ____	(13) ____	(18) ____	(23) ____
R × 1000	(4) ____	(9) ____	(14) ____	(19) ____	(24) ____
R × 10,000	(5) ____	(10) ____	(15) ____	(20) ____	(25) ____

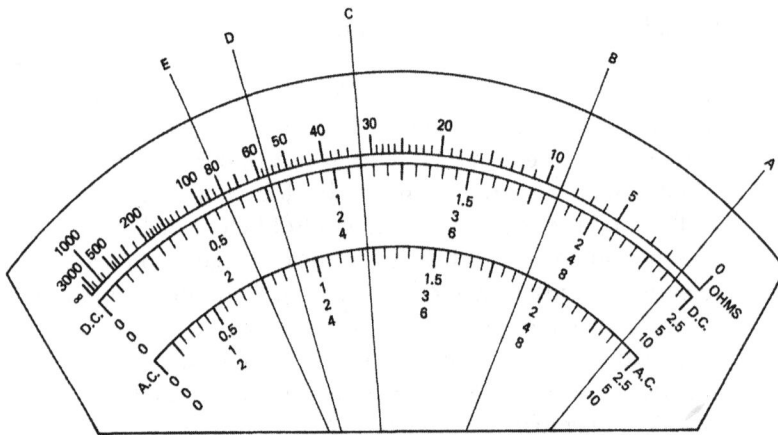

Answers

1. 1 Ω	2. 10 Ω
3. 100 Ω	4. 1 kΩ
5. 10 kΩ	6. 9 Ω
7. 90 Ω	8. 900 Ω
9. 9 kΩ	10. 90 kΩ
11. 32 Ω	12. 320 Ω
13. 3200 Ω	14. 32 kΩ
15. 320 kΩ	16. 56 Ω
17. 560 Ω	18. 5600 Ω
19. 56 kΩ	20. 560 kΩ
21. 80 Ω	22. 800 Ω
23. 8000 Ω	24. 80 kΩ
25. 800 kΩ	

Voltage is applied to electric equipment to cause it to operate. It is important to be able to measure voltage to check the operation of equipment. Many electric problems develop because either too much or too little voltage is applied to the equipment. A voltmeter is used to measure voltage in an electric circuit. A VOM also can be used to measure voltage. In the controls of the VOM in Fig. 2-2 the voltage ranges are 3, 12, 60, 300, 1200, and 6000 V. When the function-select switch is adjusted to 3 V on the dc volts range, the meter measures up to 3 V. The same is true of the other ranges of dc voltage. The voltage value of each range is the maximum value of voltage that may be measured with the VOM set on that range.

When making voltage measurements, adjust the function-select switch to the highest range of dc voltage. Connect the red and black test leads to the meter by putting them into the proper jacks. The red test lead should be put into the jack labeled V-Ω-A. The black test lead should be put into the jack labeled –COM.

It is easy to become familiar with the part of the meter scale that is used to measure dc voltage. Refer to the VOM scale of Fig. 2-3. The part of the scale below the ohms scale is the dc voltage scale. This scale usually is black. There are three dc voltage scales: 0 to 12 V, 0 to 60 V, and 0 to 300 V. All dc voltages are measured with one of these scales. Each of the dc voltage ranges on the function-select switch corresponds to a number on the right side of the meter scale or a number that can be easily multiplied or divided by 10 to equal the number on the function switch.

When the 12, 60, or 300 V range is used, the scale is read directly. In these ranges, the number to which the needle points is the actual value of the voltage being measured. When the 3, 1200, or 6000 V range is used, the number to which the needle points must be multiplied or divided. If the meter needle points to the number 50 while the meter is adjusted to the 60 V range, the measured voltage is 50 V. If the needle points to the number 250 while the meter is adjusted to the 3 V range, the measured voltage is 2.5 V ($250 \div 100 = 2.5$). When the 1200 V range is used, the numbers on the 0 V to 12 V scale are read and then multiplied by 100. Most VOMs have several scales. Some of these scales are read directly, whereas others require multiplication or division.

Before making any measurements, choose the proper dc voltage range. The value of the range being used is the maximum value of voltage that can be measured on that range. For example, when the range selected is 12 V, the maximum voltage the meter can measure is 12 V. Any voltage higher than 12 V could damage the meter. To measure an unknown voltage (no indication of its value), start by using the highest range on the meter. Then slowly adjust the range downward until a voltage reading is indicated on the right side of the meter scale.

Matching meter polarity to voltage polarity is important in the measurement of dc voltage. The meter needle moves backward, possibly damaging to the meter, if polarities are not connected properly. Meter polarity is simple to determine. The positive (+), red, test lead is connected to the positive side of the dc voltage being measured. The negative (–), black, test lead is connected to the negative side of the dc voltage being measured. The meter is always connected across (in parallel with) the dc voltage being measured.

How to Measure DC Voltage with a VOM

Voltage is the electric pressure that causes current to flow in a circuit. A common voltage source is a battery. Batteries come in many sizes and voltage values. The voltage applied to a component determines how much current will flow through it.

A dc voltmeter or the dc voltage ranges of a VOM are used to measure dc voltage. The upper left part of the VOM function-select switch of Fig. 2-2 is used for measuring dc voltage. The dc voltage ranges are 3, 12, 60, 300, 1200, and 6000 V.

When measuring voltage with a VOM, first put the test leads into the meter. The red test lead is plugged into the hole marked volts-ohms-amps (V-Ω-A). The black test lead is plugged into the hole marked negative common (–COM). The scale of the meter is used to indicate voltage (volts). The left side of the dc voltage range is marked zero and the right side is marked 300, 60, and 12. The meter needle rests on the zero until a voltage is measured. These three scales are used to measure dc voltages on the sample meter scale.

To measure a dc voltage, select the proper range. The ranges for measuring dc voltage are on the upper left part of the function-select switch. If the range setting is on the 3 V, the voltage being measured cannot be larger than 3 V. If the voltage is greater than 3 V, the meter would probably be damaged. You must be careful to use a meter range that is larger than the voltage being measured. Each of the dc voltage ranges on the function-select switch corresponds to a number on the right side of the meter scale or a number that can be easily multiplied or divided to equal the number on the function-select switch. When the 12, 60, or 300 V range is used, the dc voltage scale is read directly. On these ranges, the number to which the needle points is the actual value of the voltage being measured. When the 3, 1200, or 6000 V range is used, the number to which the needle points is multiplied or divided by 100. If the meter needle points to the number 850 while the 3 V range is being used, the measured voltage is 8.5, because 850 ÷ 100 = 8.5.

Examples of dc voltage measurements with the meter set on the 3 V range follow. If the test leads of the meter are placed across a voltage source and the meter needle moves to point A on the scale of Fig. 2-6, the dc voltage is equal to 100 ÷ 100, or 1 V. The reading at point B is 165 ÷ 100, or 1.65 V. At point C, the reading is 280 ÷ 100, or 2.8 V. There is some difficulty in reading the voltage divisions on the scales. Look at the division marks from 200 to 250. The difference between 200 and 250 is 50 units (250 – 200 = 50). There are 10 division marks between 200 and 250. The voltage per division mark is 50 ÷ 10, or 5 V, per division. So each division mark between 200 and 250 equals 5 V. This procedure is like reading a ruler or other types of scales.

If the range switch is changed to the 12 V position, the voltage is read directly from the meter scale. For example, if

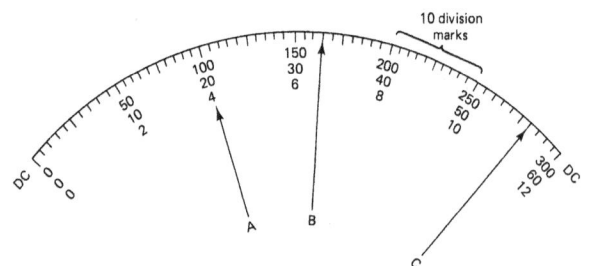

FIGURE 2-6 DC voltage scale of a VOM.

the range is set on 12 V and the meter needle moves to point A in Fig. 2-6, the voltage is 4 V. The reading at point B equals 6.6 V. At point C, the reading is 11.2 V. The same procedure is used for all other ranges.

When measuring voltage, always be sure to select the proper range. The range used is the maximum value of voltage that can be measured on that range. For example, when the range selected is 12 V, the maximum voltage that the meter can measure is 12 V. Any voltage greater than 12 V could damage the meter. When measuring an unknown voltage, start with the highest range setting on the meter. Then slowly adjust the range setting to lower values until the meter needle moves to somewhere between the center and right side of the meter scale.

In measurement of dc voltage, polarity is important. The proper matching of meter polarity and voltage source polarity must be assured. The negative (black) test lead of the meter is connected to the negative polarity of the voltage being measured. The positive (red) test lead is connected to the positive polarity of the voltage. If the polarities are reversed, the meter needle will move backward and the meter might be damaged.

A certain amount of voltage is needed to cause electric current to flow through a resistance in a circuit. The voltage is called *voltage drop.* Voltage drop is measured across any component through which current flows. The polarity of a voltage drop depends on the direction of current flow. Current flows from the negative polarity of a battery to the positive polarity. In Fig. 2-7, the bottom of each resistor is negative. The top of each resistor is positive. The negative test lead of the meter is connected to the bottom of the resistor. The positive test lead is connected to the top. The meters are connected as shown to measure each of the voltage drops in the circuit. If the meter polarity were reversed, the meter needle would move in the wrong direction.

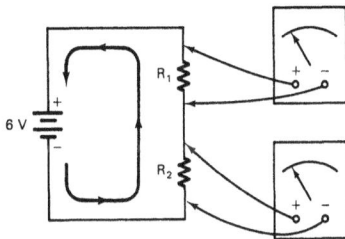

FIGURE 2-7 Measuring voltage drop in a dc circuit.

Measuring Current

Current flows through a complete electric circuit when voltage is applied. Many important tests are made by means of measuring current flow in electric circuits. The current values in an electric circuit depend on the amount of resistance in the circuit. Learning to use an ammeter to measure current in an electric circuit is important.

Most VOMs can be used to measure dc current. Refer to the controls of the VOM shown in Fig. 2-2. The function-select switch may be adjusted to any of five ranges of direct current, 12 A, 120 mA, 12 mA, 1.2 mA, and 60 μA. For example, when the function-select switch is placed in the 120 mA range, the meter is capable of measuring up to 120 mA of current. The value of the current set on the range is the maximum value that can be measured on that range. The function-select switch should first be adjusted to the highest range of direct current. Current is measured by connecting the meter into a circuit, as shown in Fig. 2-8. This is referred to as connecting the meter in *series* with the circuit.

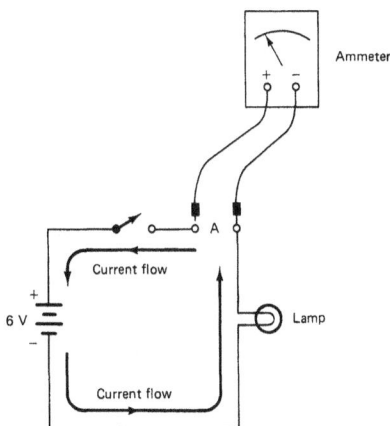

FIGURE 2-8 Meter connection for measuring dc current.

Current flows from a voltage source when a device that has resistance is connected to the source. When a lamp is connected to a battery, a current flows from the battery through the lamp. In the circuit of Fig. 2-8, electrons flow from the negative battery terminal, through the lamp, and back to the positive battery terminal. Electrons are so small that the human eye cannot see them, but their movement can be measured with an ammeter.

As the voltage applied to a circuit increases, the current increases. If 12 V is applied to the lamp in Fig. 2-8, a larger current flows through the lamp. If 24 V is applied to the same lamp, an even larger current flows. As resistance gets smaller, current increases. Resistance is the opposition to current flow. When a circuit has more resistance, it has less current flow.

How to Measure DC Current with a VOM

Refer to the dc current ranges of the VOM shown in Fig. 2-2. The ranges begin with 12 A. The next ranges are for measuring 120 mA, 12 mA, 1.2 mA, and 60 μA. There are a total of five current ranges. The function-select switch is adjusted to any of these five ranges for measuring dc. When measuring current, make it a habit to start with the meter set on its highest range. Then move the range setting to a lower value if the meter needle moves only a small amount. The most accurate reading is obtained when the meter needle is between the center of the scale and the right side. The same scales on the VOM often are used for measuring dc and dc voltage.

If the meter range is set on the 12 A range, the scale is read directly. The bottom dc scale, which has the number 12 on the right side, is used. Some examples are shown in Fig. 2-9 with the meter set on the 12 A range. At point A on the scale the reading is 4.6 A. The reading at point B is 8.8 A.

The 60 μA range on the meter is for measuring very small currents. Measurements in this range are read directly from the meter scale. The number 60 is the middle number on the right of the dc scale.

When the meter is set on the 120 mA range, the meter measures up to 120 mA of direct current. The readings on the scale are multiplied by 10 on this range setting. The readings at the points shown in Fig. 2-9 for the 120 mA range are as follows:

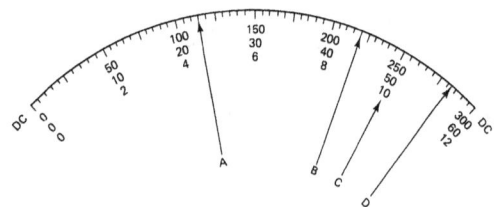

FIGURE 2-9 DC current scale of a VOM.

$$\text{Point } A = 46 \ (4.6 \times 10) \text{ mA}$$

$$\text{Point } B = 88 \ (8.8 \times 10) \text{ mA}$$

$$\text{Point } C = 100 \ (10 \times 10) \text{ mA}$$

$$\text{Point } D = 113 \ (11.3 \times 10) \text{ mA}$$

The reading at point D is halfway between the 11.2 and 11.4 divisions on the scale, so the reading is 11.3×10, or 113 mA.

The test lead polarity of the VOM is important for measuring direct current. The VOM is connected to allow current to flow through the meter in the right direction. The negative test lead is connected nearest to the negative side of the voltage source. The meter is then connected into the circuit. To measure current, a wire is removed from the circuit to place the meter into the circuit. No voltage should be applied to the circuit while the current meter is being connected. The meter is placed in series with the circuit. Series circuits have one path for current flow.

The proper procedure for measuring current through point *A* in the circuit of Fig. 2-8 is as follows:

1. Turn off the voltage source of the circuit by opening the switch.

2. Set the meter to the highest current range (12 A).

3. Remove the wire at point A.

4. Connect the negative test lead of the meter to the negative side of the voltage source.

5. Connect the positive test lead to the end of the wire that was removed from point A.

6. Turn on the switch to apply voltage to the circuit.

7. Look at the meter needle to see how far it has moved up the scale.

8. Adjust the meter range until the needle moves to between the center of the scale and the right side.

Always remember the following safety tips when measuring current with a VOM:

1. Turn off the voltage before connecting the meter to prevent electric shock. This is an important habit to develop. Always remember to turn off the voltage before connecting the meter.

2. Set the meter to its highest current range. This assures that the meter needle does not move too far to the right of the scale and possibly damage the meter.

3. A wire is disconnected from the circuit and the meter is put in series with the circuit. Always remember to disconnect a wire and reconnect the wire to one of the meter test leads. If a wire is not removed to put the meter into the circuit, the meter will not be connected properly.

4. Use the proper meter polarity. The negative test lead is connected so that it is nearest the negative side of the voltage source. The positive test lead is connected so that it is nearest the positive side of the voltage source.

Self-Examination

Refer to Fig. 2-10 and fill in the blanks with the values that correspond to the pointer location.

DC volts range	A	B	C	D	E
2.5 V dc	(26) _____	(32) _____	(38) _____	(44) _____	(50) _____
10 V dc	(27) _____	(33) _____	(39) _____	(45) _____	(51) _____
50 V dc	(28) _____	(34) _____	(40 _____	(46) _____	(52) _____
250 V dc	(29) _____	(35) _____	(41) _____	(47) _____	(53) _____
500 V dc	(30) _____	(36) _____	(42) _____	(48) _____	(54 _____
1000 V dc	(31) _____	(37) _____	(43) _____	(49) _____	(55 _____

DC amps range	A	B	C	D	E
10 A	(56) _____	(64) _____	(72) _____	(80) _____	(88) _____
2.5 A	(57) _____	(65) _____	(73) _____	(81) _____	(89) _____
500 mA	(58) _____	(66) _____	(74) _____	(82) _____	(90 _____
100 mA	(59) _____	(67 _____	(75) _____	(83) _____	(91) _____
50 mA	(60) _____	(68 _____	(76 _____	(84) _____	(92) _____
10 mA	(61) _____	(69 _____	(77) _____	(85) _____	(93) _____
2.5 mA	(62) _____	(70) _____	(78) _____	(86) _____	(94) _____
250 μA	(63) _____	(71) _____	(79) _____	(87) _____	(95) _____

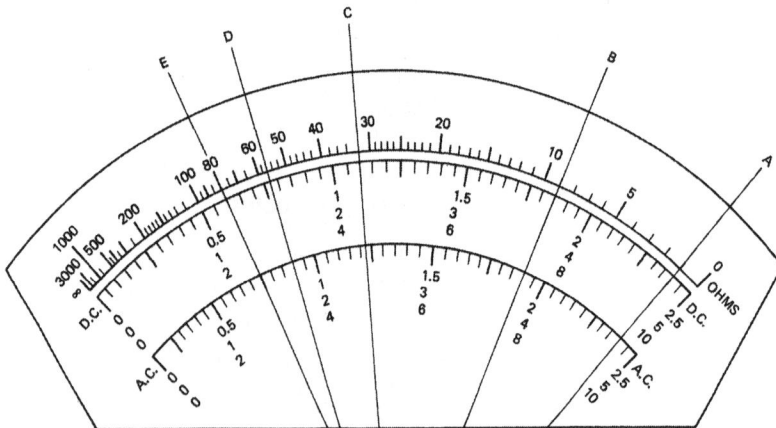

Figure 2-10

26. 2.4 V	27. 9.6 V
28. 48 V	29. 240 V
30. 480 V	31. 960 V
32. 1.85 V	33. 7.4 V
34. 37 V	35. 185 V
36. 370 V	37. 740 V
38. 1.1 V	39. 4.4 V
40. 24 V	41. 110 V
42. 240 V	43. 440 V
44. 0.755 V	45. 3.1 V
46. 15.5 V	47. 75.5 V
48. 155 V	49. 310 V
50. 0.6 V	51. 2.4 V
52. 12 V	53. 60 V
54. 120 V	55. 240 V
56. 9.6 A	57. 2.4 A
58. 480 mA	59. 96 mA
60. 48 mA	61. 9.6 mA
62. 2.4 mA	63. 240 µA
64. 7.4 A	65. 2.4 A
66. 370 mA	67. 74 mA
68. 37 mA	69. 7.4 mA
70. 1.85 mA	71. 185 µA
72. 4.4 A	73. 1.1 A
74. 210 mA	75. 44 mA
76. 22 mA	77. 4.4 mA
78. 1.1 mA	79. 110 µA
80. 3.1 A	81. 0.755 A
82. 155 mA	83. 31 mA
84. 15.5 mA	85. 3.1 mA
86. 0.755 mA	87. 75.5 µA
88. 2.4 A	89. 0.6 A
90. 120 mA	91. 24 mA
92. 12 mA	93. 2.4 mA
94. 0.6 mA	95. 60 µA

Parallel Circuit Measurements

To measure current through path 1 in the parallel circuit of Fig. 2-11, use the following procedure:

1. Open the switch to make sure no voltage is applied to the circuit.

2. Remove wires 1 and 2 from point A. Set the meter to the highest current range.

4. Connect wires 1 and 2 to the positive test lead of the meter.

5. Connect the negative test lead of the meter to point A.

6. Turn on the switch to apply voltage to the circuit.

7. Adjust the meter, if necessary, to a lower range to obtain an accurate reading.

8. Read the current value on the scale of the meter.

To measure the resistance of a parallel circuit, first remove the voltage source. Prepare the meter to measure resistance. Be sure to zero the meter. Connect the meter across the points where the circuit was connected to the voltage source (points A and B in Fig. 2-11a). Adjust the meter range, if necessary, to get an accurate resistance reading. Be sure to zero the meter each time a change of ranges is made. Once the proper range is selected, accurately read the resistance on the meter scale.

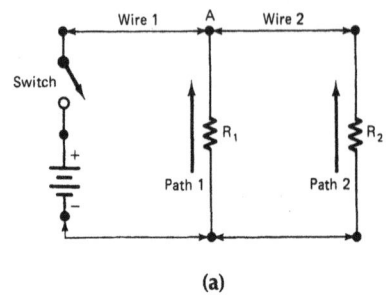

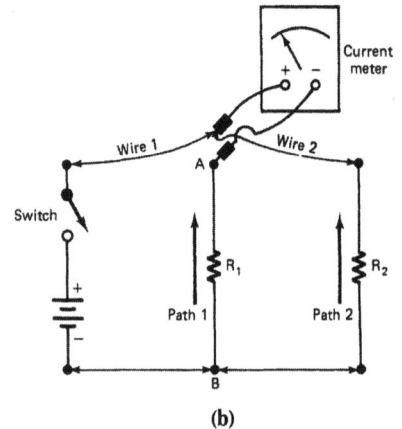

(a)

(b)

FIGURE 2-11 Making measurements in a parallel circuit. (a) Original circuit. (b) Circuit set up to measure current through path 1.

Combination Circuit Measurements

To measure the total resistance of the circuit of Fig. 2-12, first remove the voltage source from the circuit. Prepare the meter to measure resistance. Be sure to zero the meter. Connect the meter across points A and C. These points are where the voltage source was connected into the circuit. Adjust the meter range if necessary to obtain an accurate resistance reading. Once the proper range is selected, read the measured resistance on the meter scale.

To measure the current through R_1, first make sure that no voltage is applied to the circuit. Remove the wire at point C. Set the VOM to the highest current range. Connect the negative VOM test lead to point C and the positive test lead to the positive power source terminal. Now apply voltage to the circuit. Adjust the meter, if necessary, to a lower range to obtain an accurate current reading. Then read the current value on the scale of the meter. The current through R_1 should equal 2.5 A. This value is the same as the total current (I_T) of the circuit because R_1 is a series resistor.

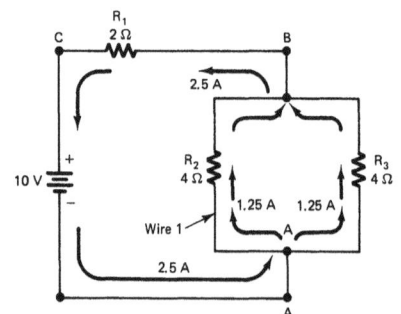

FIGURE 2-12 Combination circuit.

To measure the current through R_2 in this combination circuit, use the same procedure as for any parallel path. The following procedure should be followed:

1. Make sure that no voltage is applied to the circuit.

2. Remove wire 1 from point *A*.

3. Set the meter to its highest current range.

4. Connect wire 1 to the positive test lead of the meter.

5. Connect the negative test lead of the meter to point *A*.

6. Apply voltage to the circuit.

7. Adjust the meter, if necessary, to a lower current range to obtain an accurate reading.

8. Read the current value on the scale of the meter. The current through R_2 should be 1.25 A.

To measure the voltage across R_1, connect the negative meter lead to point *B* and the positive lead to point *C*. The voltage should equal 5 V. The voltage across R_2 equals the voltage across R_3, because they are in parallel. The negative meter lead is connected to point *A* and the positive lead to point *B*. The voltage across R_2 and R_3 should be 5 V.

Digital Meters

Many digital meters are now in use. They have numerical read-outs to simplify measurement and to make the measurements more accurate. Instruments such as digital counters, digital multimeters, and digital voltmeters are commonly used. Digital meters such as the ones shown in Fig. 2-13 rely on the operation of digital circuitry to produce a numerical readout of the measured quantity.

The readout of a digital meter is designed to transform electric signals into numerical data. Both letter and number readouts are available, as are seven-segment, discrete-number, and bar-matrix displays. Each method has a device designed to change electric energy into light energy on the display.

(a)

(b)

FIGURE 2-13 (a) A 3½-digit multimeter used for electric measurement—resistance, voltage, and current. (b) Seven-segment display used with digital multimeters (DMMs).

Self-Examination

Use the meter you will be using for completing experiments. Study the meter and answer each of the following questions. Either an analog or digital meter may be used. If the self-examination item does not apply to your meter, place N/A in the blank.

96. What company manufactured the meter?
   _____

97. What is the model number of the meter?
   _____

98. The meter will measure up to _____ A of dc.

99. Alternating current is read on the _____ colored scales.

100. True or False: The ohms-adjust control is used each time the resistance range is changed.
   _____

101. To measure dc greater than 1 A, the range switch is placed in the _____ position.

102. For measuring current in a circuit, the meter should be connected in (series or parallel)
   _____.

103. For measuring voltage, the meter should be connected in (series or parallel) _____.

104. To measure 18 mA of current, the range switch should be placed in the _____ position.

105. To measure 10 μA of current, the range switch should be placed in the _____ range.

106. To measure resistance, the red test lead must be placed in the jack marked _____ and the black test lead in the jack marked _____.

107. The most accurate resistance reading is located on the (right or left) side of the meter scale.
   _____

108. True or False: Polarity must be observed when measuring ac voltage. _____

109. True or False: Polarity is not important when measuring resistance. _____

110. Up to _____ V can be measured with the meter.

111. For measuring a resistor valued at 10 Ω, the _____ range should be used.

112. True or False: Polarity must be observed when measuring dc. _____

113. True or False: It is correct to measure the resistance of a circuit with voltage applied.
_____

114. True or False: When measuring an unknown value of voltage, one should start at the highest scale and work down to the correct scale.
_____

115. True or False: Meters should be handled with care and safety. _____

Answers

96.	*	97.	*
98.	*	99.	*
100.	True	101.	*
102.	Series	103.	Parallel
104.	*	105.	*
106.	*	107.	*
108.	False	109.	True
110.	*	111.	*
112.	True	113.	False
114.	True	115.	True

*Depends on the type of meter used.

EXPERIMENT 2-1

MEASURING RESISTANCE

Resistance is the term used to describe the opposition encountered by electric current flow. Resistance is measured in ohms (Ω). In unit 1, you learned how to determine the value of a resistor by examining its color bands. However, the resistance of many components cannot be determined through observation and therefore must be measured.

This lab is designed for an analog meter so that scale interpretation may be learned. Simple changes will easily allow use of a digital meter, if desired.

OBJECTIVE

To learn how to measure resistance with the ohmmeter portion of a multimeter.

EQUIPMENT

Multimeter (VOM)

Potentiometer: 5 kΩ

Resistors: 10 Ω, 15 Ω, 220 Ω, 470 Ω, 1 kΩ, 5.1 kΩ, 68 kΩ, 100 kΩ, 220 kΩ, 1 MΩ

PROCEDURE

1. The multimeter, or VOM, is the meter most often used for basic electronics work. A VOM can be used as an ammeter, a voltmeter, or an ohmmeter. Simply adjust the function-select switch for the desired function. Fig. 2-2 shows the controls, including the function-select switch, of a common VOM.

The function-select switch is in the center of some circular divisions and acts as a dial. A portion of the circular divisions is designated the *ohms function*. The area of the divisions located within the ohms-function space is divided into four positions: ×1, ×100, ×1000, and ×100,000. The meter you are using may be slightly different from the one illustrated. Nevertheless, it exhibits the same basic characteristics as the VOM shown. Adjust the function-select switch on your meter to the ohms function. List the different positions in the ohms-function area on your meter. _____

2. Now that you have adjusted the VOM to the ohms function, connect the test leads to the meter. The test leads used with the VOM usually are red and black. The colors are used to help you differentiate the positive and negative polarities of the meter. For our purposes red indicates the positive polarity (+), and black indicates the negative (–) polarity. Look at the meter controls in Fig. 2-2. In your VOM insert the red test lead into the hole or jack marked with (+) and the

black test lead into the hole or jack marked with (–). (*Note:* The meter you are using may be different from the one illustrated. If this is so, plug the red and black test leads into the appropriate jacks and proceed.) Touch the test leads together. Describe what happens to the needle located on the scale of the meter. _____

3. You have now adjusted the function-select switch to the ohms function of the VOM, inserted the proper test leads into the proper jacks, and seen that when the test leads are touched together, or shorted, the needle of the meter moves from the extreme left-hand side of the meter scale to the extreme right-hand side. If you are using a DVM, the display will indicate *0*. When the leads are not touched together, the digit *1* will appear on the left display in most cases. The *1* indicates infinite resistance.

4. You should now become familiar with the scale of the meter. Fig. 2-3 is a scale of a very commonly used VOM. If you are using a DVM, skip this section.

The top scale (from 0 to ∞, labeled *ohms*) is used for measuring ohms only. On most VOMs, the top scale is designated as the ohms scale. Record the location and color of the ohms scale on your meter.

_____	_____
Location	Color

5. To measure any resistance in ohms, you must select the proper position or range. On the meter illustrated in Fig. 2-2, there are four ranges: ×1, ×10, ×1000, and ×100,000. The ohmmeter must be properly zeroed before any attempt is made to accurately measure resistance. To zero the ohmmeter, short the two test leads together. This should cause the needle to move from infinity (∞) on the left to zero (0) on the right. Infinity represents a very *high* resistance, whereas zero represents a very *low* resistance. If the needle does not reach zero or if it goes past zero when the test leads are shorted, the control marked *ohms adjust* must be used to adjust the needle to rest on zero when the test leads are touched together. There will be a similar adjustment on your VOM if yours is different from the one illustrated. Adjust the VOM to each of its ranges: ×1,

×10, ×1000, and ×100,000 and zero the meter for each of these ranges. The ohmmeter must be zeroed before each resistance measurement and after each range change. Otherwise your measurements will be incorrect.

6. The ohms scale on the meter is *nonlinear.* The divisions on the right side of the scale are farther apart than those on the left side. This provides a more accurate measurement of resistance when the needle of the meter deflects and stops somewhere between the center of the ohms scale and zero.

Choosing the proper range adjustment controls where the needle deflects. Adjust the function-select switch to the ×1 range of the ohms function. Zero the meter, and then connect it across a 10 Ω resistor. Record the precise resistance (in ohms) indicated by the needle on the ohms scale in Fig. 2-1A. The range selected for this operation was ×1, which means that the number to which the needle points will be multiplied by 1.

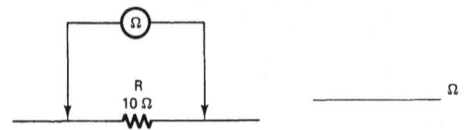

R
10 Ω

_____ Ω

FIGURE 2-1A

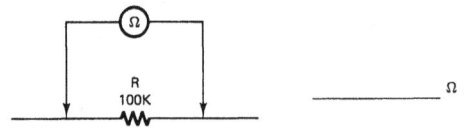

R
100K

_____ Ω

FIGURE 2-1B

7. Adjust the function-select switch to the ×100,000 range. Zero the meter and connect it to a 100 kΩ resistor, as illustrated in Fig. 2-1B. Record the resistance (in ohms) shown by the needle on the ohms scale. The range selected for this measurement is ×100,000, which means that the number to which the needle points will be multiplied by 100,000.

8. Measure and record the values of the resistors indicated in Fig. 2-1C. Remember to choose a meter range that will cause the needle to deflect somewhere between the center of the scale and zero. Always zero the meter when changing ranges, and always multiply the number indicated on the scale by the multiplier of the chosen range: ×1, ×100, ×1000, ×100,000. Never measure the resistance of a component until it has been disconnected.

9. Using the proper procedure for measuring resistance, measure and record the precise resistance of the 200 Ω pot illustrated in Fig. 2-1D.

10. Adjust the control of the pot while the ohmmeter is connected. Describe how this action affects the measured resistance of the pot. _____

11. Alter the connections of the pot as illustrated in Fig. 2-1E.

12. Adjust the pot both clockwise and counterclockwise and describe how this action affects its measured resistance. _____

Color-coded value	Measured value in ohms
15 Ω	
100 Ω	
220 Ω	
470 Ω	
1 kΩ	
5.1 kΩ	
68 kΩ	
220 kΩ	
1 MΩ	

FIGURE 2-1C

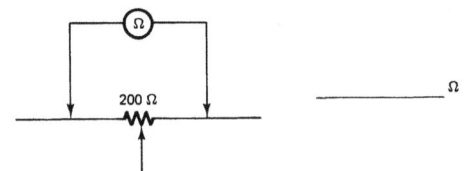

200 Ω

_____ Ω

FIGURE 2-1D

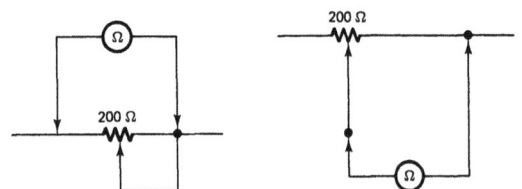

200 Ω

200 Ω

FIGURE 2-1E

ANALYSIS

1. Why is the ohms scale of an analog multimeter considered to be nonlinear? _____

2. Where on an analog ohms scale are the most accurate measurements found? _____

3. What is meant by the ×1000 range on the ohm-meter? _____

4. What is meant by *zeroing* the ohmmeter?

   _____

5. Why is it necessary to zero the ohmmeter?

   _____

6. If the range of the ohmmeter was set to ×100,000 and the needle pointed to 0.6 on the ohms scale, what would be the value of the resistance being measured? _____

EXPERIMENT 2-2

MEASURING VOLTAGE

Voltage is the force that causes electric current to flow through conductors or paths. This force, sometimes called *electromotive force* (EMF), is measured in volts. It is essential that anyone involved in electronics work be able to use the necessary equipment to measure voltage accurately.

This lab is designed for an analog meter so that scale interpretation may be learned. Simple changes allow the use of a digital meter, if desired.

OBJECTIVES

1. To learn to measure voltage with a multimeter.

2. To construct basic electric circuits.

EQUIPMENT

Multimeter (VOM)

Potentiometer: 5 kΩ

Resistors: 470 Ω, 1 kΩ

6 V battery or power supply

6 V lamp with socket

Connecting wires

PROCEDURE

1. You learned in Experiment 2-1 that a multimeter can be used to measure resistance. A multimeter also can be used to measure voltage. Experiment 2-1 showed that the function-select switch can be adjusted to cause the multimeter to perform many measurement functions with several different ranges. The dc voltage ranges of the meter shown in Fig. 2-2 are 3, 12, 60, 300, 1200, and 6000 V. When the function-select switch is adjusted to 3 V on the dc volts range, the meter can be used to measure a *maximum* of 3 V. The same is true of the remaining ranges within the dc volts function. The numerical value of the chosen range indicates the maximum value of voltage that may be measured on that range. The meter you are using may differ somewhat from the VOM illustrated in Fig. 2-2. Nevertheless, it will exhibit the same basic characteristics as the one described.

2. Adjust the function-select switch on your meter to the lowest range of the dc volts function. List the different ranges within the dc volts function on your meter.

—————, —————, —————,

—————, —————

VOM dc voltage range	VOM scale	Multiplier	Divisor
3 V	0 to 300	–	100
12 V	0 to 12	None	None
60 V	0 to 60	None	None
300 V	0 to 300	None	None
1200 V	0 to 12	100	–
6000 V	0 to 60	100	–

FIGURE 2-2A VOM scales, ranges, and multipliers.

3. Connect the red and black test leads to the meter by inserting the appropriate ends into the proper jacks on the face of the meter. The red test lead should be inserted into the jack labeled (+) and the black test lead should be inserted into the jack labeled –COM. (*Note:* The meter you are using may be different. If this is the case, plug the red and black test leads into the appropriate jacks and proceed.)

4. You now have the VOM properly equipped and adjusted to measure dc volts. You should now become familiar with the portion of the scale of the meter used for measuring dc volts. Figure 2-3 shows the scale of a very commonly used VOM. The scale immediately under the ohms scale is the dc volts scale. There are three dc volts scales: 0 to 12 V, 0 to 60 V, and 0 to 300 V. All dc voltages are measured with one of these scales. The question becomes, What scales are to be used with what meter ranges? Each of the dc voltage ranges corresponds to a number on the right side of the meter scale or a multiple or divisor of that number. Figure 2-2A indicates the proper scale to be read for each dc voltage range.

5. When you choose the 12 V, 60 V, or 300 V range, you read the scale directly. In these ranges, the number to which the needle points is the actual value of the voltage being measured. When you select the 3 V range, the number to which the needle points must be divided by 100. For example, if the needle of the meter points to 50 while the meter is adjusted to the 3 V range, the measured voltage is 0.5 V. (*Note:* The meter you are using may differ slightly from the one described. VOMs usually have several scales. Some of these scales can be read directly, whereas others necessitate use of a multiplier or divisor. If your VOM is different, identify the scales that can be read directly and the ones that must be used with multipliers or divisors.)

6. Complete Fig. 2-2B by indicating the proper voltage scale and voltage value. Show the multiplier or divisor for the voltage range for the meter scale shown. This will give you practice using a meter scale.

7. You are now almost ready to measure dc voltage. Before actually doing so you should become familiar with some facts about choosing the dc voltage range. Recall from step 1 that the numerical value of the chosen range indicates the *maximum* value

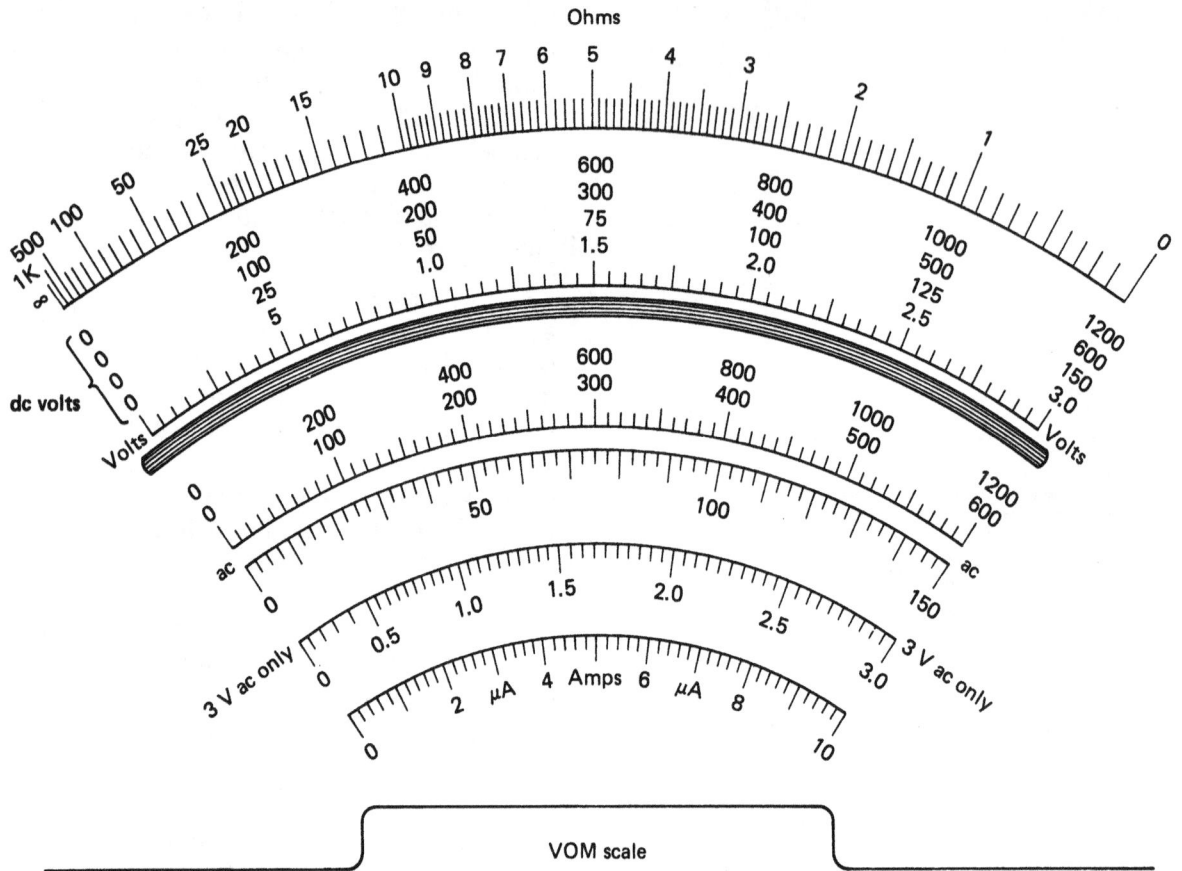

FIGURE 2-2B VOM scale and range settings.

VOM range	Proper scale to read	Number needle points to	Voltage value	Multiplier	Divisor
Example: 3 V	0 to 3	1.5	1.5	None	None
3 V		2.0			
15 V		100			
60 V		300			
150 V		75			
600 V		200			
1200 V		800			
15 V		125			
60 V		350			
150 V		125			
60 V		125			
15 V		75			

Voltage to be measured	Range to select	Scale to read	Multiplier	Divisor
Example: 2.0 V	3 V	0 to 3.0	None	None
3.5 V				
0.5 V				
6 V				
14 V				
17 V				
30 V				
65 V				
100 V				
250 V				
1100 V				
Unknown				

FIGURE 2-2C Selecting VOM ranges and scales.

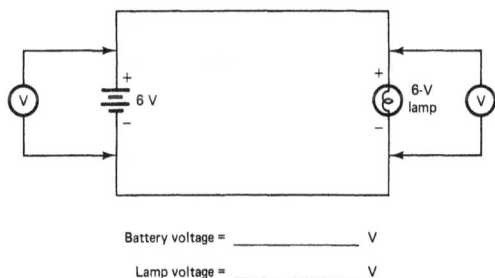

Battery voltage = _____ V

Lamp voltage = _____ V

FIGURE 2-2D

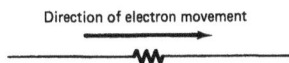

Direction of electron movement

FIGURE 2-2E

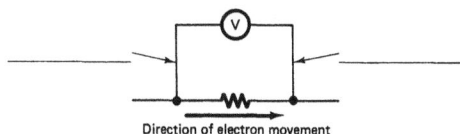

Direction of electron movement

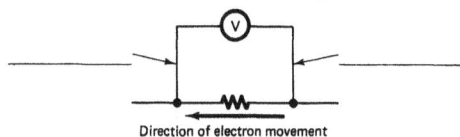

Direction of electron movement

FIGURE 2-2F

of voltage that may be measured on that range. When the range selected is 12 V, the maximum voltage that the meter can measure is 12 V. Any voltage above 12 V could damage the meter on this range. This is true for most voltage scales on most VOMs. If you are trying to measure a voltage that is totally unknown (no indications as to its approximate value), you should start the measuring procedure by choosing the *highest* range on your meter and slowly adjusting the range downward until a voltage reading is indicated on the upper or right-hand half of the meter scale. Complete Fig. 2-2C by showing the proper meter range to be selected, the proper meter scale to be read, and the proper multiplier and divisor to be used with each voltage to be measured. Use the scale of your meter to complete the figure.

8. You need to consider meter *polarity* when measuring dc voltage. Correct matching of meter polarity to voltage polarity is vital. If proper care is not exercised, the meter will deflect backward, and the internal components will be damaged. Meter polarity is simple to determine. The positive (+), red, test lead is connected to the positive side of the dc voltage to be measured. The negative (–), black, test lead is connected to the negative side of the dc voltage to be measured.

9. Measure and record the actual voltage of a 6 V battery. The voltmeter is always connected *across* the voltage, or in parallel with the voltage to be measured.

   Actual voltage of battery = _____ V

10. Construct the circuit illustrated in Fig. 2-2D. Measure and record the voltage supplied by the battery and the voltage across the lamp.

11. A certain amount of voltage is always necessary to cause current flow. This voltage is called the *voltage drop* and is found across any component through which current flows. The polarity of this voltage drop is determined by the direction of the movement of electrons from negative to positive. Thus if electrons were moving through a resistor in the direction shown in Fig. 2-2E, the left side of the resistor would be negative (–) and the right side would be positive (+). The black test lead of the meter would be connected to the left side of the resistor and the red test lead would be connected to the right side. This procedure would measure the voltage drop developed across this resistor. In the illustrations of Fig. 2-2F, indicate the proper meter polarity for measuring the voltage drop across each resistor.

12. Construct the circuit of Fig. 2-2G. Measure and record the battery voltage and the voltage drops across R_1 and R_2. Remember that electrons move from negative to positive.

13. Construct the circuit of Fig. 2-2H. Measure and record the voltage drops across R_1 and R_2.

14. Construct the circuit shown in Fig. 2-2I.

15. Adjust the potentiometer both counterclockwise and clockwise and record the voltages.

    Counterclockwise voltage = _____ V

    Clockwise voltage = _____ V

ANALYSIS

1. What is voltage?
   _____
   _____

2. If you were measuring an unknown voltage what meter range would you choose? Why?
   _____
   _____

3. What determines the polarity of a voltage drop developed by a resistor through which electrons are moving?
   _____
   _____

4. What causes the voltage drop developed across any component through which electrons are moving?
   _____
   _____

5. When is it necessary to use a multiplier or divisor with a scale on an analog multimeter?
   _____
   _____

6. How is the proper range of a multimeter selected?
   _____
   _____

7. Could 4 V be measured on the 3 V range of a multimeter? Why?
   _____
   _____

8. Indicate the proper range of your meter for measuring the following dc voltages:

   10 V = _____ range   100 V = _____ range

   2.8 V = _____ range   65 V = _____ range

   6 V = _____ range   40 V = _____ range

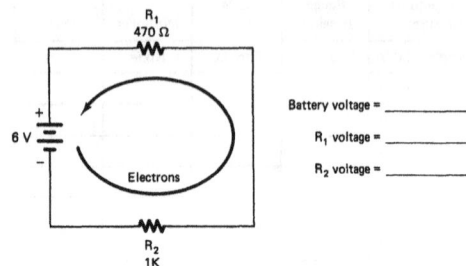

FIGURE 2-2G

Battery voltage = _____

R_1 voltage = _____

R_2 voltage = _____

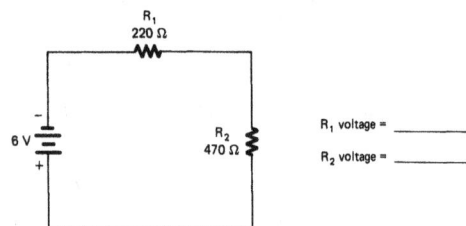

FIGURE 2-2H

R_1 voltage = _____

R_2 voltage = _____

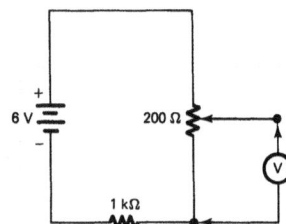

FIGURE 2-2I

EXPERIMENT 2-3

MEASURING CURRENT

Current is the movement of electrons from one location to another through conductors or paths. Current is measured in amperes (the number of electrons moving past a given point in a circuit per second), milliamperes (0.001 A), or microamperes (0.000001 A). It is essential that anyone involved in electronics work learn to measure current values accurately.

An analog meter is used in this activity so that scale interpretation can be learned. Simple changes allow the use of a digital meter, if desired.

OBJECTIVES

    1. To learn to measure current with an ammeter.

    2. To construct basic electric circuits.

    3. To convert amperes to milliamperes or microamperes and vice versa.

EQUIPMENT

    Multimeter (VOM)

    Resistors: 220 Ω, 1 kΩ, 100 kΩ, 220 kΩ

    6 V battery or power supply

    SPST switch

    Connecting wires

PROCEDURE

    1. You have learned from previous experiments that a VOM can be used to measure resistance and voltage. In addition to these electric quantities, current can be measured with a VOM. Figures 2-2 and 2-3 show the controls and the scale of a common VOM. The function-select switch controls the measurement function and the range of the VOM. Figure 2-2 indicates that the function-select switch may be adjusted to one of five ranges within the dc function, namely, 12 A, 120 mA, 12 mA, 1.2 mA, and 60 μA. Thus when the function-select switch is placed in the 120 mA position within the dc function, the meter can be used to measure a maximum of 120 mA. The other dc current ranges of the VOM are similarly arranged. The numerical value of the chosen range shows the *maximum* value of current that can be measured on that range.

    2. The meter you are using may differ somewhat from the illustrated VOM. It should, however, exhibit the basic characteristics of the one described. Adjust the function-select switch on your meter to the lowest range of the dc function. List the different ranges for measuring dc on your meter. _____, _____, _____, _____, _____

3. Plug the black and red test leads into the jacks labeled (–) and (+), respectively. (*Note:* If your meter is different from the one illustrated, plug the test leads into the appropriate jacks and proceed.)

4. You are now ready to become familiar with the portion of the scale of the meter used for measuring direct current. Figure 2-3 illustrates the scale of a commonly used VOM. Three scales are used for measuring dc. They are the same as those used to measure dc voltage. Five dc ranges can be used with this meter. All current measurements are read on these scales with the proper multiplier or divisor. Figure 2-3A shows the proper multiplier or divisor to be used with the dc scales for any of the five ranges. (*Note:* Your meter may have different scales and multipliers from those in Fig. 2-3A. Identify them and proceed.)

5. Each range of dc has a specific multiplier or divisor or is read directly on the scale. For example, if the needle of the meter points to the number 6 on the dc scale, and the meter is adjusted to the 12 mA range, the measured current is 6 mA. If the needle points to 6 while the meter is adjusted to the 120 mA range, the measured current is 60 mA ($6 \times 10 = 60$, or $6 \div 0.1 = 60$).

6. Complete Fig. 2-3B by determining the proper current value when the needle points to a specific number. Also determine the correct multiplier and divisor to be used with the indicated current range. You should use the equivalent ranges, scales, multipliers, and divisors of the meter in Figs. 2-2 and 2-3.

7. Most VOMs can be used to measure currents greater than 1 A. Up to 12 A of current may be measured with the meter illustrated. The numerical value of the chosen range indicates the *maximum* value of current that can be measured in that range. For example, when the range selected is 12 mA, the meter can be used to measure a maximum of 12 mA. Any current above 12 mA could damage the meter in this range. This is true for most current scales of most VOMs. If you are trying to measure an unknown current (no indication of its approximate value), start the measuring procedure by choosing the highest range on your meter and slowly adjusting the range downward until a current reading is indicated on the scale. Complete Fig. 2-3C by indicating the proper meter range to be selected and the proper multiplier and divisor to be used with each value of current to be measured. (*Note:* Always select

VOM current range	VOM scale	Multiplier	Divisor
60 μA	0–60	None	None
1.2 mA	0–12	–	10
12 mA	0–12	None	None
120 mA	0–12	10	–
12 A	0–12	None	None

FIGURE 2-3A VOM dc scales, ranges, and multipliers.

Equivalent range	Number needle points to	Current value	Multiplier	Divisor
Example: 12 mA	8	8 mA	None	None
120 mA	2			
120 mA	7			
1.2 mA	4			
12 mA	9			
60 μA	5			
120 mA	7.5			
1.2 mA	3.2			
60 μA	6			
12 A	3			

FIGURE 2-3B Determining VOM current values.

Current to be measured	Range to select	Multiplier	Divisor
Example: 0.9 mA	1 mA	0.1	10
2 A			
900 mA			
25 μA			
2 mA			
85 mA			
6 mA			
0.5 mA			
16 mA			
9 mA			
Unknown current			

FIGURE 2-3C Determining meter ranges.

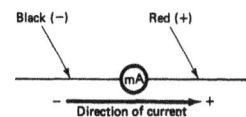

FIGURE 2-3D

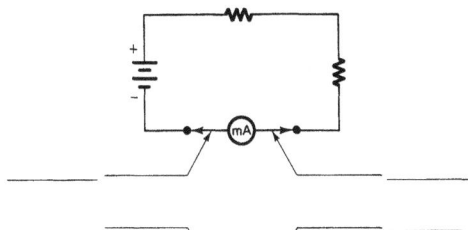

FIGURE 2-3E

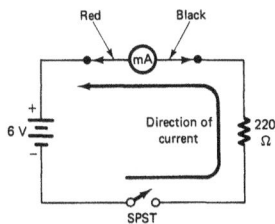

FIGURE 2-3F Connecting the VOM for measuring
dc current.

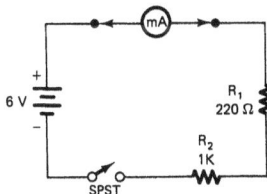

FIGURE 2-3G Adding a resistor to circuit.

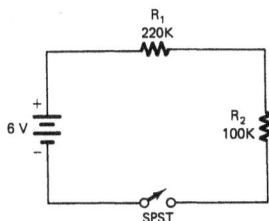

FIGURE 2-3H Low-current circuit.

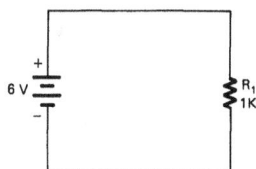

FIGURE 2-3I Simple Resistive Circuit.

the smallest range that will allow you to measure the current.) Use *your* meter's ranges and scales for completing Fig. 2-3C.

8. Meter polarity of the VOM is important when measuring current. The VOM must always be connected in such a way as to allow the current to flow through the meter in the proper direction. Otherwise the needle will deflect backward, possibly causing internal damage to the meter. If current is to flow *through* the meter in the proper direction, consideration must be given to the direction of current. Current flows from negative to positive; thus if the meter is connected correctly into a circuit, current will flow through the black test lead (–), through the meter, and then through the red test lead (+). The illustration of Fig. 2-3D should be helpful.

9. In the illustrations of Fig. 2-3E show which test leads should be red (+) and which should be black (–) for the current meters indicated.

10. Adjust the VOM to the dc function 120 mA range (or equivalent) and connect it into the circuit, as illustrated in Fig. 2-3F. Be sure to observe the proper meter polarity.

11. Close the SPST switch and record the current indicated by the VOM: _____ mA, or _____ A.

12. Open the SPST switch and add the additional resistor illustrated in Fig. 2-3G.

13. Change the range of the VOM to 12 mA (or equivalent), and close the switch. Record the current indicated on the VOM: _____ mA, or _____ A.

14. Disconnect the circuit of Fig. 2-3G and connect the circuit of Fig. 2-3H.

15. Set the VOM to the 60-µA range (or equivalent), and connect the meter into the circuit, observing proper polarity. Close the switch and record the current: _____ µA, or _____ mA, or _____ A.

16. Disconnect the circuit illustrated in Fig. 2-3H.

17. Construct the circuit illustrated in Fig. 2-3I.

18. Adjust the VOM to the 12 mA range (or equivalent) and connect it into the circuit illustrated in Fig. 2-3I, observing proper polarity. Record the current: _____ mA, or _____ A.

ANALYSIS

1. What is current?

   _____

   _____

2. What does each of the dc ranges on your multi-meter indicate?

   _____

   _____

3. Describe how a multimeter is properly connected into a circuit to measure current.

   _____

   _____

4. Why is proper polarity important to observe when measuring dc current?

   _____

   _____

5. What meter ranges would be selected on your meter to measure the following currents?

Current	Range
0.009 A	_____
0.11 A	_____
0.8 A	_____
0.5 mA	_____
0.000040 A	_____

6. What is the proper procedure to follow when attempting to use a multimeter to measure an unknown current?

   _____

   _____

7. What is the proper procedure to be followed when attempting to use a multimeter to measure a current greater than 2 A?

   _____

   _____

FIGURE 2-4A

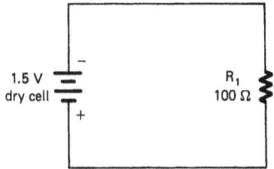

FIGURE 2-4B Simple resistive circuit.

FIGURE 2-4C Sketch of power supply.

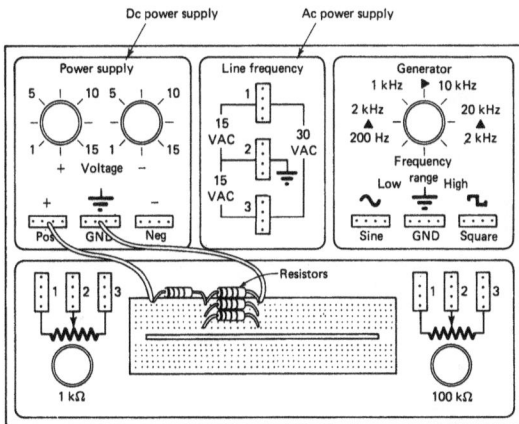

FIGURE 2-4D Common power supply.

EXPERIMENT 2-4

FAMILIARIZATION WITH POWER SUPPLY

The variable dc power supply is an electronic circuit that changes alternating current (ac) into dc. This circuit is usually housed in a case or protective enclosure. Output jacks are provided for external connections. This type of power supply usually has some means by which the output voltage can be adjusted from zero (0) to a maximum value. The symbol for the variable dc power supply is illustrated in Fig. 2-4A.

OBJECTIVE

To become familiar with the use of a power supply to provide dc voltage to an electronic circuit.

EQUIPMENT

Multimeter (VOM)

1.5 V dry cell

Variable dc power supply

Resistor: 100 Ω

PROCEDURE

1. A dry cell or battery can be used as a simple power source for a circuit. Acquire the 1.5 V dry cell and prepare the VOM to measure dc voltage. Measure and record the precise voltage across the cell: _____ V.

2. Construct the circuit illustrated in Fig. 2-4B.

3. The circuit represents a dry cell in a *loaded* state (large current output). Measure the dry cell voltage with it loaded as shown in Fig. 2-4B: _____ V.

4. How did the voltage measured in step 1 compare with the measured voltage in step 3? How do you account for any difference?

   _____

   _____

5. In Fig. 2-4C, make a simple sketch of the variable dc power supply that you will use for this and future experiments. Include in your sketch all controls, output jacks, and range switches, where applicable.

6. A common type of power supply is shown in Fig. 2-4D.

7. Adjust the VOM to measure dc voltage. Connect it to the dc output jacks of the power supply. These are usually labeled (+) and (−) or ground.

8. Turn on the power supply and adjust it to produce maximum dc voltage. Record the maximum voltage measured on the VOM.

Maximum voltage = _____ V.

9. Adjust the variable dc voltage control on the power supply to cause it to produce different voltages measured with the VOM. What is the voltage range of your power supply?

Range = _____ V to _____ V.

10. Turn off the power supply and disconnect the VOM.

11. If the power supply operating instructions are available, read them thoroughly.

ANALYSIS

1. Why is it considered correct to check the condition of a dry cell while it is loaded?

_____

_____

2. What are some advantages of using dry cells?

_____

_____

3. What are some advantages of using a variable dc power supply?

_____

_____

4. Draw the symbols for a dry cell and a variable dc power supply.

5. What is meant by the term *ground?*

_____

_____

6. Draw the symbol for a ground.

Measuring Voltage, Current, and Resistance

Instructions: For each of the following, circle the answer that most correctly completes the statement.

1. A VOM is an example of a
   a. Multifunction meter
   b. Single-function meter
   c. Galvanometer
   d. High-resistance meter

2. For measuring voltage, the meter is connected in
   a. Series
   b. Parallel
   c. Series-parallel
   d. Either series or parallel

3. For measuring current flow, the meter is connected in
   a. Series
   b. Parallel
   c. Series-parallel
   d. Either series or parallel

4. The ohmmeter setting that would be best when checking the continuity of a fuse is:
   a. R × 1
   b. R × 10
   c. R × 1000
   d. R × 100,000

5. The current indicated on the analog meter scale of Figure 2E-5 is
   a. 5 A
   b. 25 A
   c. 5 mA
   d. 25 mA

6. When measuring resistance with an ohmmeter, a short indicates
   a. Infinite resistance
   b. Zero resistance
   c. Midscale resistance
   d. Cannot measure a short with an ohmmeter

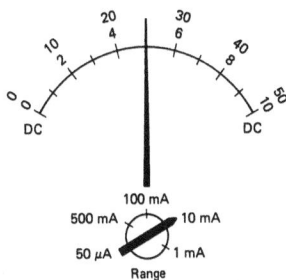

FIGURE 2E-5

7. The current indicated on the meter scale of Figure 2E-7 is
   a. 100mA
   b. 200 mA
   c. 10 mA
   d. 2 mA

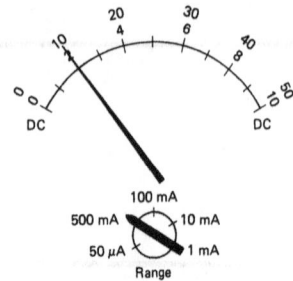

FIGURE 2E-7

8. The current indicated on the analog meter scale of Figure 2E-8 is
   a. 40 A
   b. 8 A
   c. 40 mA
   d. 8 mA

9. The current indicated on the analog meter scale of Figure 2E-9 is
   a. 5 mA
   c. 25 μA
   d. 25 mA

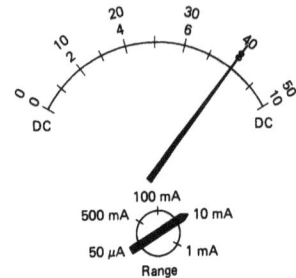

FIGURE 2E-8

10. A major precaution to observe when measuring resistance with an ohmmeter is to make sure
    a. Circuit power is turned off
    b. Circuit power is turned on
    c. Range switch is set on maximum setting
    d. Polarity is observed

True-False: Place either T or F in each blank.

_____ 11. When using a multimeter to measure dc voltage, polarity should be observed.

_____ 12. *Ohms adjust* should be used each time the resistance range is changed.

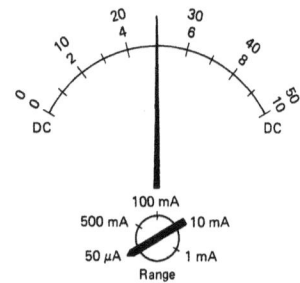

FIGURE 2E-9

_____ 13. For measuring current flow, the meter should be connected in series.

_____ 14. For measuring volts, the meter should be connected in parallel.

_____ 15. The most accurate ohms reading of an analog multimeter is found on the left side of the ohms scale.

_____ 16. Polarity is important when measuring ac volts.

_____ 17. Polarity is not important when measuring resistance.

_____ 18. For measuring a resistor valued at 10 Ω, the ×1 scale of an ohmmeter should be read.

_____ 19. It is considered good practice to measure the resistance of a circuit with voltage present in the circuit.

_____ 20. When measuring an unknown value of dc voltage, one should start at the highest range and work down to the correct range.

Ohm's Law and Electric Circuits

To understand electronics, it is necessary to know how to apply basic electric theory. Electronics is a somewhat mathematical subject. The mathematics is easy to understand, because it has practical applications that are easy to see. The basic theory used is called *Ohm's law*. Learn Ohm's law; it applies to the basic theory of electric circuits. All the examples in this unit are dc circuits. Alternating current (ac) circuits are more complex. AC circuits are described in a companion text, *Understanding AC Circuits.*

UNIT OBJECTIVES

Upon completion of this unit, you will be able to do the following:

1. Solve series, parallel, and series-parallel (combination) circuits using Ohm's law.

2. Design voltage-divider circuits.

3. Analyze bridge circuits.

4. Explain the operation of a Wheatstone bridge.

5. Analyze circuits using Kirchhoff's voltage and current laws.

6. List problem-solving methods useful in solving complex circuit problems.

7. Solve basic math problems with a calculator.

8. Define Ohm's law and the power equation.

9. Solve problems finding current, voltage, and resistance.

10. Calculate power using the proper power formulas.

11. Explain maximum power transfer.

12. Define voltage drop in a circuit.

13. Solve circuit problems with resistors in different configurations.

Important Terms

Review the following terms before studying unit 3.

Branch A path of a parallel circuit.

Branch current The current through a parallel branch.

Branch resistance The total resistance of a parallel branch.

Branch voltage The voltage across a parallel branch.

Bridge circuit A circuit with four resistors in a special arrangement.

Circuit A path through which electric current flows.

Combination circuit A circuit that has one portion connected in series with the voltage source and another part connected in parallel.

Complex circuit Another term for a combination circuit.

Current The flow of electric charges through a circuit path or conductor.

Difference in potential The voltage across two points of a circuit.

Direct proportion The condition in which an increase or decrease in one quantity causes an increase or decrease in another quantity.

Equivalent resistance A resistance value that is the same value in a circuit as two or more parallel resistances in the circuit.

Inverse The value of 1 divided by some quantity, such as $1/R$ for finding parallel resistance.

Inverse proportional The condition in which an increase or decrease in one quantity causes another quantity to do the opposite.

Kilowatt-hour (kWh) 1000 watts per hour, a unit of measurement of electric energy.

Kirchhoff's current law The principle by which the sum of the current flowing into any point or junction of conductors of a circuit is equal to the sum of the currents flowing away from that point.

Kirchhoff's voltage law The principle by which in any current loop of a circuit, the sum of the voltage drops is equal to the voltage supplied to that loop; with proper signs (− or +), the algebraic sum of the voltage sources and voltage drops in a circuit is equal to zero.

Maximum power transfer A condition that exists when the resistance or impedance of a load (R_L) equals that of the source that supplies it (R_s).

Ohm's law The law that explains the relation of voltage, current, and resistance in electric circuits.

Parallel circuit A circuit that has two or more current paths.

Power (_P_) The rate of doing work in electric circuits, found with the equation $P = I \times V$.

Reciprocal The same as _inverse._

Resistance Opposition to the flow of current in an electric circuit.

Series circuit A circuit that has one path for current flow.

Superposition A problem-solving method for electronic circuits.

Total current The current that flows from the voltage source of a circuit.

Total resistance The total opposition to current flow of a circuit; found by removing the voltage source and connecting an ohm-meter across the points where the source was connected.

Total voltage The voltage supplied by a source.

Voltage Electric pressure that causes current to flow through a circuit.

Voltage drop The voltage across two points of a circuit; found with the equation $V = I \times R$.

Watt (W) The unit of electric power.

Use of Calculators

Electronic calculators greatly simplify problem solving for electronic technology. Low-cost calculators have the mathematical functions necessary to solve most problems encountered by electronics technology students. Most calculators have a simplified instruction manual that explains how to accomplish each mathematical function. The four basic functions (+, −, ×, ÷) are used for many simple electronics problems. Other functions that are particularly helpful in electronics applications are listed in the following table:

Function	Key	Operation
Exponent	EE or EXP	Raises a number to a power of 10
Inverse	1/x	Divides 1 by the number entered
Square	x²	Multiplies a number by itself (e.g., $5^2 = 5 \times 5$)
Square root	√X	Extracts the square root of a number
Trigonometric functions (sine, cosine, tangent)	sin cos tan	Determines the trigonometric function of an angular value (e.g., sin 30° = 0.5)
Inverse trig functions	inv sin inv cos inv tan	Determines the angular value of a trigonometric function (e.g., inv sin 0.5 = 30°)
Logarithms	log	Finds the logarithm of a number (e.g., log 150 = 2.17)
Inverse logarithms	inv log	Finds number when logarithm value is given (e.g., inv log 2.176 = 150)
Storage	sto	Stores a number in memory
Recall	rcl	Recalls a number from memory

Ohm's Law

Ohm's law is the most basic and most used of all electric theories. Ohm's law explains the relation of voltage (the force that causes current to flow), current (the movement of electrons), and resistance (the opposition to current flow). Ohm's law is stated as follows: *An increase in voltage increases current if resistance remains the same.* Ohm's law stated in another way is: An increase in resistance causes a decrease in current if voltage remains the same. The electric values used with Ohm's law usually are represented with capital letters. For example, voltage is represented by the letter *V,* current by the letter *I,* and resistance by the letter *R.* The mathematical relation of the three electric units is shown in the following formulas. Memorize these formulas. The Ohm's law circle of Fig. 3-1 is helpful in remembering the formulas.

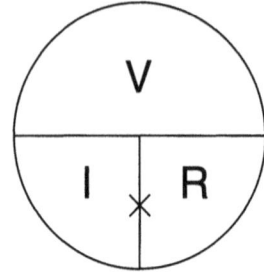

FIGURE 3-1 Ohm's law circle. *V,* voltage; *I,* current; *R,* resistance. To use the circle, cover the value you want to find and read the other values as they appear in the formula: *V = I × R, I = V/R, R = V/I.*

$$V = I \times R$$

$$I = \frac{V}{R}$$

$$R = \frac{V}{I}$$

Voltage (*V*) is measured in volts. Current (*I*) is measured in amperes. Resistance (*R*) is measured in ohms. If two electric values are known, the third value can be calculated with one of the formulas. Look at Fig. 3-2. Using the Ohm's law current formula, *I = V/R,* the calculated value of *I* in the circuit is

$$I = \frac{V}{R} = \frac{10V}{10\Omega} = 1A$$

If the voltage in the circuit is doubled, as shown in Fig. 3-3, the calculated current using the Ohm's law current formula, *I = V/R,* is as follows:

$$I = \frac{V}{R} = \frac{20V}{10\Omega} = 2A$$

In this example as voltage is increased, current increases if resistance remains the same. If voltage is doubled, current doubles. If voltage is increased 10 times, current increases 10 times. Now look at Fig. 3-4a. The current flow in the circuit is calculated as follows:

$$I = \frac{V}{R} = \frac{10V}{100\Omega} = 0.1A$$

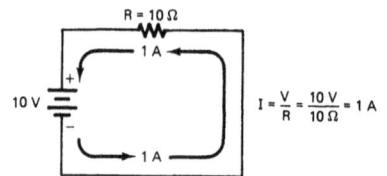

FIGURE 3-2 Example of Ohm's law.

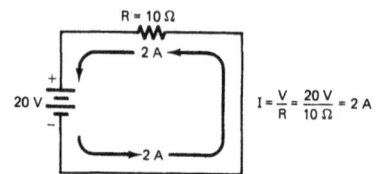

FIGURE 3-3 Effect of doubling voltage.

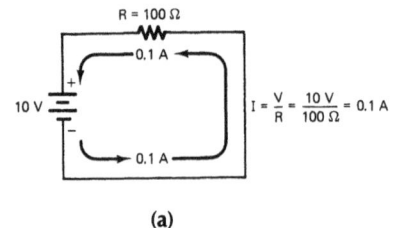

(a)

FIGURE 3-4 Effect of increasing resistance.

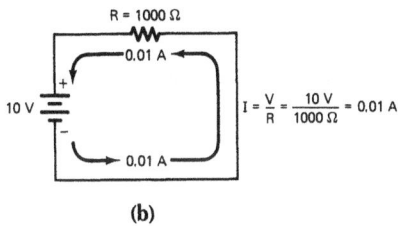

(b)

FIGURE 3-4 Effect of increasing resistance.

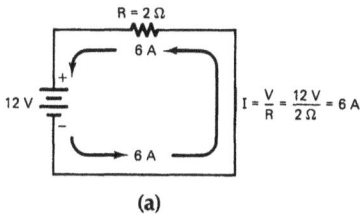

(a)

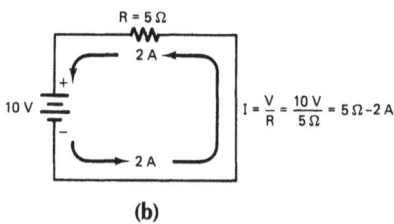

(b)

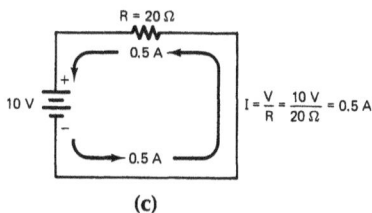

(c)

FIGURE 3-5 Examples of Ohm's law.

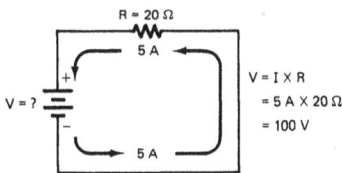

FIGURE 3-6 Use of Ohm's law to find voltage.

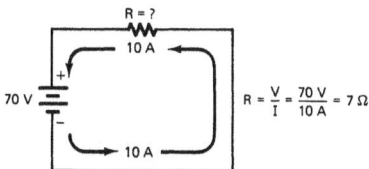

FIGURE 3-7 Use of Ohm's law to find resistance.

In Fig. 3-4b, the 100 Ω resistor in the circuit is replaced with a 1 kΩ resistor. The value of the 10 V current applied to the 1 kΩ resistance is calculated as follows:

$$ I = \frac{V}{R} = \frac{10V}{1000\Omega} = 0.01A $$

As resistance increases, current decreases.

Ohm's law explains the relation of voltage, current, and resistance in electric circuits. The circle shown in Fig. 3-1 is used for remembering this relation. To calculate the voltage in a circuit, cover the V on the circle: V equals I times R. To find current, cover the I: I equals V over R. To find resistance, cover the R: R equals V over I. The circle helps in using Ohm's law to solve simple electric problems.

Another example of using Ohm's law is shown in Fig. 3-5a. To find the value of current that flows in this circuit, use the Ohm's law circle. Cover the I and find that I equals V over R. In this circuit, I = 12 V divided by 2 Ω. So the current is 6 A.

Another example of Ohm's law is shown in Fig. 3-5b. The voltage of this circuit is 10 V and the current is 2 A. The resistance of the circuit is 10 V divided by 2 A, or 5 Ω. Ohm's law also states that if the resistance of a circuit increases, the current decreases if the voltage stays the same. An example of this is shown in Fig. 3-5c. The resistance of the circuit is increased to 20 Ω. The current of the circuit is now 10 V divided by 20 Ω, or 0.5 A. The current in the previous circuit was 2 A. If the resistance of a circuit is increased four times, the current flow decreases to one-fourth its original value.

The previous examples deal mainly with calculating current. Voltage and resistance of a circuit also can be calculated with Ohm's law. In a circuit in which current and resistance are known, Ohm's law is used to find voltage. In the circuit of Fig. 3-6, assume that resistance is 20 Ω and the current is 5 A. Use the circle and cover V. This shows that V equals I times R. So the voltage required to cause 5 A of current through a 20Ω resistance is 5 A times 20 Ω, or 100 V.

Ohm's law also is used to find the value of resistance in a circuit. Assume that a circuit has a known value of voltage and current. The resistance required to cause this value of current flow can be calculated. In the example in Fig. 3-7, the voltage is 70 V and the current equals 10 A. The resistance of the circuit is found with the Ohm's law circle. Cover the R and find that R is equal to V over I. So the resistance of this circuit is equal to 70 V divided by 10 A, or 7 Ω.

Self-Examination

Use Ohm's law to solve the following problems.

1. A doorbell requires 0.5 A of current to ring. The voltage applied to the bell is 120 V. What is its resistance? _____ Ω.

2. A relay used to control a motor has a 50 Ω resistance. It draws a current of 0.5 A. What voltage is required to operate the relay? _____ V.

3. An automobile battery supplies a current of 10 A to the starter. It has a resistance of 1.25 Ω. What is the voltage delivered by the battery? _____ V.

4. What voltage is needed to light a lamp if the current required is 3 A and the resistance of the lamp is 80 Ω? _____ V.

5. If the resistance of a stereo circuit is 200 Ω and it draws 0.6 A, what voltage is needed? _____ V.

6. A television draws 0.25 A. The operating voltage is 120 V. What is the resistance of the TV circuit? _____ Ω.

7. The resistance of the motor of a vacuum cleaner is 30 Ω. For a voltage of 120 V, find the current. _____ A.

8. The magnet of a speaker carries 0.1 A when connected to a 50 V supply. Find its resistance. _____ Ω.

9. How much current is drawn from a 12 V battery when operating an automobile horn of 16 Ω resistance? _____ A.

10. How much current is drawn by a 2600 Ω clock that plugs into a 120 V outlet? _____ A.

Use Ohm's law to solve the following problems.

11. $I = 2$ A, $R = 500$ Ω, _____ V

12. $R = 20$ Ω, $I = 3$ A, _____ V

13. $I = 1.2$ A, $R = 1000$ Ω, _____ V

14. $R = 1.5$ kΩ, $I = 130$ mA, _____ V

15. $R = 120$ kΩ, $I = 20$ mA, _____ V

16. $V = 2$ V, $R = 1.5$ kΩ, $I =$ _____ mA

17. $R = 1.8$ kΩ, $V = 10$ V, $I =$ _____ mA

18. $R = 6.8$ kΩ, $V = 15$ V, $I =$ _____ mA

19. $R = 5.6$ kΩ, $V = 12$ V, $I =$ _____ mA

20. $V = 240\text{ V}$, $R = 1\text{ M}\Omega$, $I = $ _____ mA

21. $V = 20\text{ V}$, $I = 5\text{ A}$, $R = $ _____ Ω

22. $V = 35\text{ V}$, $I = 5\text{ mA}$, $R = $ _____ $\text{k}\Omega$

23. $I = 10\text{ mA}$, $V = 120\text{ V}$, $R = $ _____ $\text{M}\Omega$

24. $I = 260\text{ mA}$, $V = 120\text{ V}$, $R = $ _____ Ω

25. $V = 5\text{ V}$, $I = 20\text{ mA}$, $R = $ _____ $\text{k}\Omega$

Answers

1.	240 Ω	2.	25 V
3.	12.5 V	4.	240 V
5.	120 V	6.	480 V
7.	4 A	8.	500 Ω
9.	0.75 A	10.	0.046 A
11.	1000 V	12.	60 V
13.	1200 V	14.	195 V
15.	2400 V	16.	1.33 mA
17.	5.55 mA	18.	2.2 mA
19.	2.14 mA	20.	0.24 mA
21.	4 Ω	22.	7 kΩ
23.	12 MΩ	24.	461.54 Ω
25.	250 kΩ		

Series Electric Circuits

There are three types of electric circuits, *series circuits, parallel circuits,* and *combination circuits.* The easiest type of circuit to understand is the series circuit. Series circuits are different from other types of electric circuits. It is important to remember the characteristics of a series circuit.

In a series circuit, there is only one path for current to follow. Because there is only one current path, the current flow is the same value in any part of the circuit. The voltages in the circuit depend on the resistance of the components in the circuit. When a series circuit is opened, there is no path for current flow. Thus the circuit does not operate.

In the circuit examples that follow, many variables with subscripts are used, such as R_T, V_T, and I_1. It is common to use subscripts to identify electric components in circuit diagrams. The circuit shown in Fig. 3-8 has two resistors and a battery. The resistors are labeled R_1 and R_2. The subscripts identify each of these resistors. Subscripts also aid in making measurements. The voltage drop across resistor R_1 is called voltage drop V_1. The term *total* is represented by the subscript T, as in V_T. V_T is total voltage applied to a circuit. The current measurement I_2 is the current through resistor R_2 measured at point B. Total current (I_T) is measured at point A. The voltage drop across R_2 is called V_2.

Subscripts are valuable in troubleshooting and repair of electronic equipment. It would be impossible to isolate problems in equipment if components could not be easily identified.

The main characteristic of a series circuit is that it has only one path for current flow. In the circuit shown in Fig. 3-8, current flows from the negative side of the voltage source, through resistor R_1, through resistor R_2, and then to the positive side of the voltage source. Another characteristic of a series circuit is that the current is the same everywhere in the circuit. A wire at point A could be removed and a multimeter placed into the circuit to measure current. The value of current at point A is the same as the current at point B and point C of the circuit.

In any series circuit, the sum of the voltage drops is equal to the voltage applied to the circuit. The circuit shown in Fig. 3-8 has voltage drops of 12 V plus 8 V, which is equal to 20 V. Another characteristic of a series circuit is that its total resistance to current flow is equal to the sum of all resistance in the circuit. In the circuit shown in Fig. 3-9, the total resistance of the circuit is the sum of the two resistances. So the total resistance is equal to 20 Ω plus 10 Ω, or 30 Ω.

When a series circuit is opened, there is no longer a path for current flow. The circuit will not operate. In the circuit in Fig. 3-10, if lamp L_1 burns out, its filament is open. Because a series circuit has only one current path, that path is broken. No current flows in the circuit. Lamp L_2 will not work either. If one light burns out, the others go out also. This is because the series current path is opened.

Ohm's law is used to explain how a series circuit operates. In the circuit of Fig. 3-11, the total resistance is 2 Ω plus 3 Ω, or 5 Ω. The applied voltage is 10 V. Current is equal to voltage divided by resistance, or $I = V/R$. In the circuit shown, current is equal to 10 V divided by 5 Ω, which is 2 A. If a current meter were connected to this circuit, the current measurement would be 2 A. Voltage drops across each of the resistors also may be found. Voltage equals current times resistance ($V = I \times R$). The voltage drop across R_1 (V_1) is equal to the current through R_1 (2 A) times the value of R_1 (2 Ω). The voltage drop across R_1 equals 2 A × 2 Ω, or 4 V. The voltage drop across R_2 (V_2) equals 2 A × 3 Ω, or 6 V. The sum of these voltage drops is equal to the applied voltage. To check these values, add 4 V plus 6 V, which is equal to 10 V.

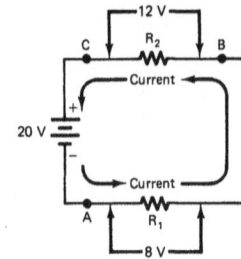

FIGURE 3-8 Series electric circuit.

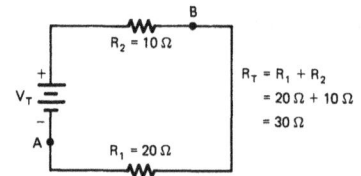

FIGURE 3-9 Finding total resistance in a series circuit.

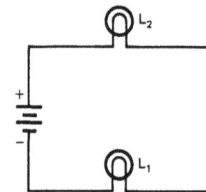

FIGURE 3-10 Two lamps connected in series.

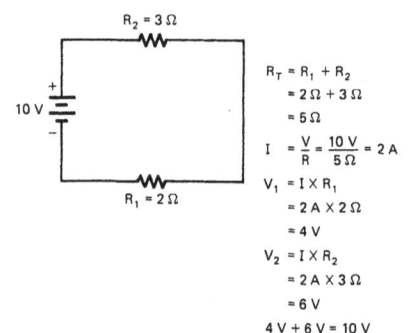

FIGURE 3-11 Use of Ohm's law for a series circuit.

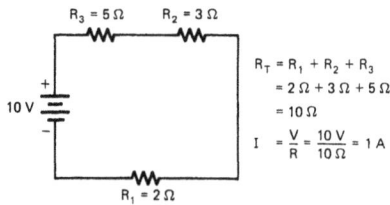

$R_3 = 5\,\Omega$ $R_2 = 3\,\Omega$

$10\,V$

$R_1 = 2\,\Omega$

$R_T = R_1 + R_2 + R_3$
$= 2\,\Omega + 3\,\Omega + 5\,\Omega$
$= 10\,\Omega$

$I = \dfrac{V}{R} = \dfrac{10\,V}{10\,\Omega} = 1\,A$

FIGURE 3-12 Effect of adding resistance to a series circuit.

If another resistance is added to a series circuit, as shown in Fig. 3-12, resistance increases. Because there is more resistance, the current flow decreases. The circuit now has R_3 (a 5 Ω resistor) added in series to R_1 and R_2. The total resistance is now 2 Ω + 3 Ω + 5 Ω, or 10 Ω, compared with 5 Ω in the previous example. The current is now 1 A, compared with 2 A in the other circuit:

$$I = \frac{V}{R} = I \ \text{or} \ = \frac{10V}{10\Omega} = 1.\text{A}$$

Summary of Series Circuits

There are several important characteristics of series circuits. Remember the following basic rules for series circuits:

1. The same current flows through each part of a series circuit.

2. The total resistance of a series circuit is equal to the sum of the individual resistances.

3. The voltage applied to a series circuit is equal to the sum of the individual voltage drops.

4. The voltage drop across a resistor in a series circuit is directly proportional to the size of the resistor.

5. If the circuit is broken at any point, no current flows.

Examples of Series Circuits

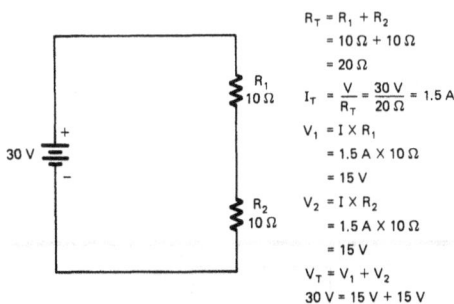

R_1
$10\,\Omega$

$30\,V$

R_2
$10\,\Omega$

$R_T = R_1 + R_2$
$= 10\,\Omega + 10\,\Omega$
$= 20\,\Omega$

$I_T = \dfrac{V}{R_T} = \dfrac{30\,V}{20\,\Omega} = 1.5\,A$

$V_1 = I \times R_1$
$= 1.5\,A \times 10\,\Omega$
$= 15\,V$

$V_2 = I \times R_2$
$= 1.5\,A \times 10\,\Omega$
$= 15\,V$

$V_T = V_1 + V_2$
$30\,V = 15\,V + 15\,V$

FIGURE 3-13 Example of a series circuit.

There is only one path for current flow in a series circuit. The same current flows through each part of the circuit. To find the current through a series circuit, only one value must be found. *Kirchhoff's current law* states that the sum of the voltage drops across the parts of a series circuit equals the applied voltage. In a series circuit, the sum of the voltage drops always equals the source voltage. This is shown in Fig. 3-13. In this circuit, a voltage (V_T) of 30 V is applied to a series circuit that has two 10 Ω resistors. The total resistance (R_T) of the circuit is equal to the sum of the resistor values (20 Ω). Ohm's law is used to find the total current of the circuit, as follows:

$$I_T = \frac{V_T}{R_T} = \frac{30V}{20\Omega} = 1.5A$$

Each resistor is 10 Ω, and the current through them is 1.5 A. The voltage drop across the resistors is then found. The voltage (V_1) across resistor R_1 is found as follows

$$V_1 = I \times R_1 = 1.5\,A \times 10\,\Omega = 15\,V$$

Resistor R_2 has the same value as resistor R_1, and the same current flows through it. The voltage drop across resistor R_2 also equals 15 V. Adding these two 15 V drops gives a total voltage of 30 V. The sum of the voltage drops is equal to the applied voltage (15 V + 15 V = 30 V).

The series circuit of Fig. 3-14 has three resistors, with values of 10 Ω, 20 Ω, and 30 Ω. The applied voltage (V_T) can be found because the current through the circuit is known to be 1 A. The circuit is a series circuit, so the same current flows through each resistor. Ohm's law ($V = I \times R$) can be used to find the voltage drop across each resistor, as follows:

$$V_1 = I \times R_1 = 1\,A \times 10\,\Omega = 10\,V$$

$$V_2 = I \times R_2 = 1\,A \times 20\,\Omega = 20\,V$$

$$V_3 = I \times R_3 = 1\,A \times 30\,\Omega = 30\,V$$

When the individual voltage drops are known, they may be added to find the applied voltage (V_T):

$$V_T = V_1 + V_2 + V_3$$

$$= 10\,V + 20\,V + 30\,V$$

$$= 60\,V$$

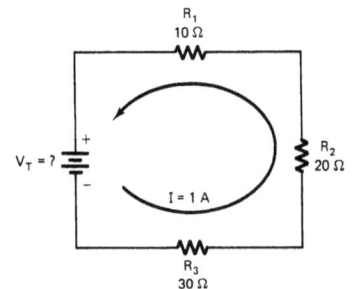

FIGURE 3-14 Series-circuit example.

The voltage drops in a series circuit are in direct proportion (when one increases, the other does too) to the resistance across which they appear. This is true because the same current flows through each resistor. The larger the value of the resistor, the larger the voltage drop across it will be.

Self-Examination

Solve each of the following series circuit problems. Find:

26. Current (I) = _____ A

27. Voltage across R (V_R) = _____ V

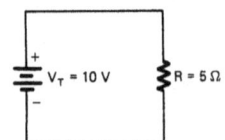

FIGURE 3SE-1

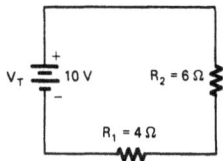
FIGURE 3SE-2

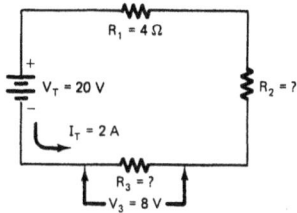
FIGURE 3SE-3

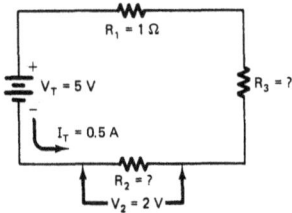
FIGURE 3SE-4

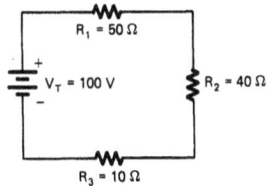
FIGURE 3SE-5

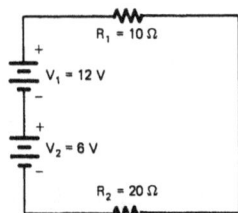
FIGURE 3SE-6

Find:

28. Total resistance (R_T) = _____ Ω

29. Current through R_1 (I_1) = _____ A

30. Current through R_2 (I_2) = _____ A

31. Voltage across R_2 (V_2) = _____ V

Find:

32. Voltage across R_1 (V_1) = _____ V

33. Resistance of R_2 = _____ Ω

34. Voltage across R_2 (V_2) = _____ V

35. Resistance of R_3 = _____ Ω

Find:

36. V_1 = _____ V

37. R_2 = _____ Ω

38. V_3 = _____ V

39. R_3 = _____ Ω

40. I_3 = _____ A

Find:

41. I_T = _____ A

42. V_1 = _____ V

43. V_2 = _____ V

44. V_3 = _____ V

45. Power converted by R_1 (P_1) = _____ W

46. Power converted by R_1 (P_2) = _____ W

47. Power converted by R_3 (P_3) = _____ W

48. Total power converted by circuit (P_T) = _____ W

Find:

49. V_T = _____ V

50. R_T = _____ Ω

51. V_1 = _____ V

52. I_T = _____ A

53. Total power (P_T) = _____ W

Answers

26.	2 A	27.	10 V
28.	10 Ω	29.	1 A
30.	1 A	31.	6 V
32.	8 V	33.	2 Ω
34.	4 V	35.	4 Ω
36.	0.5 V	37.	4 Ω
38.	2.5 V	39.	5 Ω
40.	0.5 A	41.	1 A
42.	50 V	43.	40 V
44.	10 V	45.	50 W
46.	40 W	47.	10 W
48.	100 W	49.	18 V
50.	30 Ω	51.	6 V
52.	0.6 A	53.	10.8 W

Parallel Electric Circuits

Parallel circuits are different from series circuits in several ways. A parallel circuit has two or more paths for current to flow from the voltage source. In Fig. 3-15, path 1 is from the negative side of the voltage source, through R_1, and back to the positive side of the voltage source. Path 2 is from the negative side of the voltage source, through R_2, and back to the positive side of the voltage source. Path 3 is from the negative side of the power supply, through R_3, and back to the voltage source.

In a parallel circuit, the voltage is the same across every component of the circuit. In Fig. 3-15, the voltage across points A and B is 10 V. This is the value of the voltage applied to the circuit. Following point A to point C shows that these two points are connected. Point B and point D also are connected. The voltage from point A to point B is the same as the voltage from point C to point D.

Another characteristic of a parallel circuit is that the sum of the currents through each path equals the total current that flows from the voltage source. In Fig. 3-16 the currents through the paths are 1 A, 2 A, and 3 A. One ampere of current could be measured through R_1 at point A or point B. Two amperes of current could be measured through R_2, and 3 A could be measured through R_3. The total current is 6 A. This value of total current can be measured at point C or point D in the circuit. The current is the same in every part of a series circuit, and current divides through each branch in a parallel circuit. More resistance causes

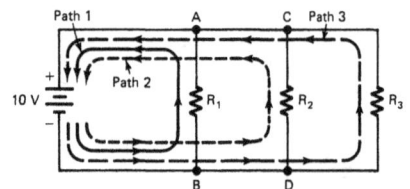

FIGURE 3-15 Parallel electric circuit.

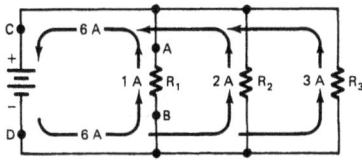

FIGURE 3-16 Current flow in a parallel circuit.

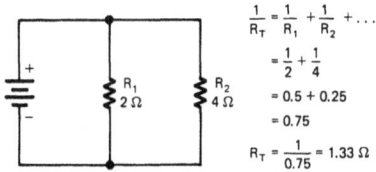

$$\frac{1}{R_T} = \frac{1}{R_1} + \frac{1}{R_2} + \ldots$$

$$= \frac{1}{2} + \frac{1}{4}$$

$$= 0.5 + 0.25$$

$$= 0.75$$

$$R_T = \frac{1}{0.75} = 1.33\ \Omega$$

or

when there are only *two* resistances:

$$R_T = \frac{R_1 \times R_2}{R_1 + R_2} = \frac{2\,\Omega \times 4\,\Omega}{2\,\Omega + 4\,\Omega} = \frac{8}{6} = 1.33\ \Omega$$

FIGURE 3-17 Finding total resistance of a parallel circuit.

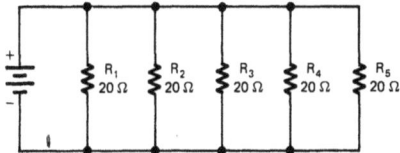

$$R_T = \frac{\text{resistance of each}}{\text{number of paths}}$$

$$= \frac{20\,\Omega}{5}$$

$$= 4\,\Omega$$

FIGURE 3-18 Finding total resistance when all resistances are the same.

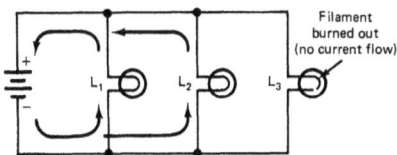

FIGURE 3-19 Three lamps connected in parallel.

less current to flow through a parallel branch. A branch is a parallel path through a circuit.

The *total resistance* (R_T) of a parallel circuit is more difficult to calculate than that of a series circuit. The formula used is: 1 divided by total resistance ($1/R_T$) is equal to $1/R_1 + 1/R_2 + 1/R_3 + 1$ divided by each of as many other resistances as there are in the circuit. This is called an inverse, or reciprocal, formula. Refer to the example of Fig. 3-17. When trying to find the total resistance of a parallel circuit, first write the formula. Each value is divided by 1. Next divide each resistance value by 1 and write the values. Add these values to get a value of 1 divided by total resistance. Do not forget to divide the value obtained by 1 to find the total resistance of the circuit. The total resistance in a parallel circuit is less than any individual resistance in the circuit. In the example shown, 1.33 Ω is less than 2 Ω or 4 Ω.

If there are *only two paths* in a parallel circuit, it is easier to find the total resistance. The following formula can be used:

$$R_T = \frac{R_1 \times R_2}{R_1 + R_2}$$

The two resistances are multiplied and then added. The multiple of the two is in the numerator of the formula, and the sum is in the denominator. When using this formula, it is not necessary to divide the resistance values by 1. Remember that this formula may be used *only* when there are two resistances in a parallel circuit. If there are more than two resistances, the reciprocal formula must be used.

Another simple method of finding total resistance in a parallel circuit is when *all the resistance values are the same*. An example of a parallel circuit with all resistances the same is a string of lights connected in parallel. Each lamp has the same resistance. When all resistances are equal, to find total resistance divide the resistance value of each resistor by the number of paths (Fig. 3-18). If five 20 Ω resistors are connected in parallel, the total resistance is equal to 20 divided by 5, or 4 Ω.

When one of the components of a parallel path is opened, the rest of the circuit continues to operate. In a series circuit, when one component is opened, no current flows in the circuit. Because the same voltage is applied to all parts of a parallel circuit, the circuit will operate unless the path from the voltage source is broken (Fig. 3-19). If lamp L_3 has a filament burned out, this causes an open circuit. No current flows through lamp L_3. However, lamps L_1 and L_2 continue to operate. When lamps are connected in parallel, if one of these lamps burns out, the rest of them still burn.

A sample problem with a parallel circuit is shown in Fig. 3-20. This circuit has a source voltage of 10 V and resistors of 5, 10, and 20 Ω. First find the total resistance of the circuit. The total resistance of this circuit is 2.85 Ω. The total opposition to current flow of this circuit is 2.85 Ω.

Find the current that flows from the voltage source. In this circuit the total current is found by dividing the voltage by the total resistance. So, 10 V divided by 2.85 Ω gives 3.5 A of current. This means that 3.5 A of current flows from the voltage source and divides into the three paths of the circuit.

Find the current through each path of the circuit. The current from the voltage source divides into each path of the circuit. A lower resistance in a path causes more current to flow through the path. Current is equal to voltage divided by resistance. The current through each path is found by dividing the voltage (10 V) by the resistance of the path. The voltage is the same across each path. The current values are 2, 1, and 0.5 A. These are added together to equal the total current, so the total current of 3.5 A is 2 A + 1 A + 0.5 A.

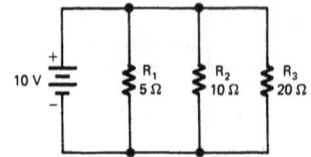

$$\frac{1}{R_T} = \frac{1}{R_1} + \frac{1}{R_2} + \frac{1}{R_3}$$
$$= \frac{1}{5} + \frac{1}{10} + \frac{1}{20}$$
$$= 0.2 + 0.1 + 0.05$$
$$= 0.35$$
$$R_T = \frac{1}{0.35} = 2.85\ \Omega$$
$$I_T = \frac{V}{R_T} = \frac{10\ V}{2.85\ \Omega} = 3.5\ A$$
$$I_1 = \frac{V}{R_1} = \frac{10\ V}{5\ \Omega} = 2\ A$$
$$I_2 = \frac{V}{R_2} = \frac{10\ V}{10\ \Omega} = 1\ A$$
$$I_3 = \frac{V}{R_3} = \frac{10\ V}{20\ \Omega} = 0.5\ A$$
$$I_T = I_1 + I_2 + I_3$$
$$= 2\ A + 1\ A + 0.5\ A$$
$$= 3.5\ A$$

FIGURE 3-20 Sample parallel circuit problem.

Summary of Parallel Circuits

Remember the following basic rules for parallel circuits:

1. There are two or more paths for current flow.

2. Voltage is the same across each component of the circuit.

3. The sum of the currents through each path is equal to the total current that flows from the source.

4. Total resistance is found with the following formula:

$$\frac{1}{R_T} = \frac{1}{R_1} + \frac{1}{R_2} + \frac{1}{R_3} + \ldots + \frac{1}{R_n}$$

5. If one of the parallel paths is broken, current continues to flow in all the other paths.

Examples of Parallel Circuits

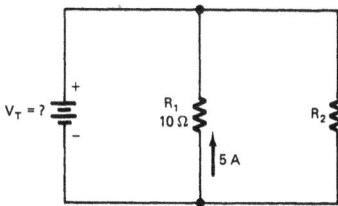

FIGURE 3-21 Example of a parallel circuit.

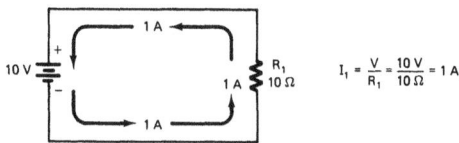

(a)

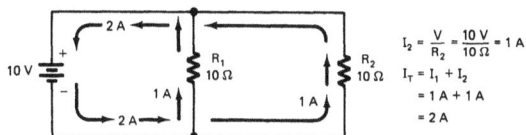

(b)

FIGURE 3-22 Current flow in a parallel circuit.
(a) One path. (b) Two paths.

Parallel circuits have more than one current path connected to the same voltage source. An example of a parallel circuit is shown in Fig. 3-21. Start at the voltage source (V_T) and look at the circuit. Two complete current paths are formed. One path is from the voltage source, through resistor R_1, and back to the voltage source. The other path is from the voltage source, through resistor R_2, and back to the voltage source. Remember that the source voltage divides across the resistors in a series circuit.

Assume that the current through resistor R_1 in Fig. 3-21 is 5 A and the value of the resistor is 10 Ω. The voltage across resistors R_1 and R_2 and the source voltage (V_T) may be found. According to Ohm's law, the voltage across resistor R_1 is as follows:

$$V_1 = I_1 \times R_1 = 5 \text{ A} \times 10 \text{ } \Omega = 50 \text{ V}$$

Because this is a parallel circuit, the voltages are the same. The source voltage (V_1) in this circuit is 50 V.

Ohm's law shows that the current in a circuit is *inversely* proportional to the resistance of the circuit. This means that when resistance increases, current decreases. The division of current in parallel circuit paths is based on this fact. The total current in a parallel circuit divides through the paths on the basis of the value of resistance in each path.

An example of the flow of current in a parallel circuit is shown in Fig. 3-22. Figure 3-22a shows a series circuit. The total current passes through one resistor (R_1). The amount of current is

$$I_1 = \frac{V_T}{R_1} = \frac{10\text{V}}{10\Omega} = 1.0\text{A}$$

Figure 3-22b shows the same circuit with another resistor (R_2) connected in parallel across the voltage source. The current through R_2 is the same as through R_1, because their resistances are equal, as follows:

$$I_2 = \frac{V_T}{R_2} = \frac{10\text{V}}{10\Omega} = 1.0\text{A}$$

Because 1 A of current flows through each of the two resistors, the total current of 2 A flows from the source. The division of currents through the resistors is shown. Each current path in the parallel circuit is called a *branch*. Each branch carries part of the total current that flows from the source.

Changing the value of any resistor in a parallel circuit has no effect on the current in the other branches. However, a change in resistance does affect total current. If R_2 of Fig. 3-22b were changed to 5 Ω (Fig. 3-22c), the total current would increase, as follows:

$$I_1 = \frac{V_T}{R_1} = \frac{10V}{10\Omega} = 1.0A$$

$$I_2 = \frac{V_2}{R_2} = \frac{10V}{5\Omega} = 2A$$

$$I_T = I_1 + I_2 = 1\,A + 2\,A = 3.0\,A$$

The total resistance of a parallel circuit is not equal to the sum of the resistors as in series circuits. When more branches are added to a parallel circuit, the total current (I_T) increases. In a parallel circuit, the total resistance is *less than any of the branch resistances.* As more parallel resistances are added, the total resistance of the circuit decreases.

There are several ways to find the total resistance of parallel circuits. The method used depends on the type of circuit. The following are common methods for finding total parallel resistance:

1. *Equal resistors.* When two or more equal-value resistors are connected in parallel, their total resistance (R_T) is as follows:

$$R_T = \frac{\text{value of one resistance}}{\text{number of paths}}$$

If four 20 Ω resistors are connected in parallel, their total resistance is 20 ÷ 4 = 5 Ω. For two 50 Ω resistors in parallel, R_T = 50 ÷ 2 = 25 Ω.

2. *Product over the sum.* Another shortcut for finding the total resistance of two parallel resistors is called the product-over-sum method, as follows:

$$R_T = \frac{R_1 \times R_2}{R_1 + R_2}$$

For example, the total resistance (R_T) of a 10 Ω resistor and a 20 Ω resistor connected in parallel is calculated as follows:

$$R_T = \frac{R_1 \times R_2}{R_1 + R_2} = \frac{10 \times 20}{10 + 20} = \frac{200}{30} = 6.67\Omega$$

The total resistance of a 10 Ω and a 20 Ω resistor is less than that of the smallest resistor. The product-over-sum method can be used only for *two* resistances in parallel.

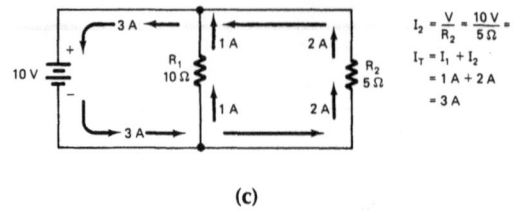

(c)

FIGURE 3-22 Current flow in a parallel circuit. (c) R_2 changed to 5 Ω.

3. *Reciprocal method.* Most circuits have more than two resistors of unequal value. The reciprocal method must then be used to find total resistance. The reciprocal method is as follows:

$$\frac{1}{R_T} = \frac{1}{R_1} + \frac{1}{R_2} + \frac{1}{R_3} + \ldots + \frac{1}{R_n}$$

For example, if 1 Ω, 2 Ω, 3 Ω, and 4 Ω resistors are connected in parallel, resistance is calculated as follows:

$$\frac{1}{R_T} = \frac{1}{1} + \frac{1}{2} + \frac{1}{3} + \frac{1}{4} = 1 + 0.5 + 0.33 + .025$$

$$= 2.08$$

$$R_T = \frac{1}{208}$$

$$= 0.48\Omega$$

Self-Examination

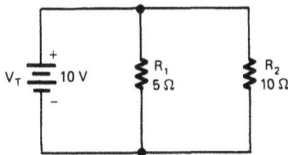

FIGURE 3SE-7

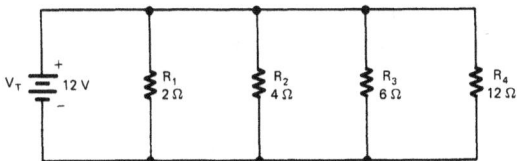

FIGURE 3SE-8

Solve each of the following parallel circuit problems.
Find:

54. Total resistance (R_T) = _____ Ω

55. Total current (I_T) = _____ A

56. Current through resistor R_1 (I_1) = _____ A

57. Current through resistor R_2 (I_2) = _____ A

Find:

58. Total resistance (R_T) = _____ Ω

59. Total current (I_T) = _____ A

60. Current through resistor R_1 (I_1) = _____ A

61. Current through resistor R_2 (I_2) = _____ A

62. Current through resistor R_2 (I_3) = _____ A

63. Current through resistor R_4 (I_4) = _____ A

64. Total power (P_T) = _____ W

Find:

65. Total resistance (R_T) = _____ Ω

66. Total current (I_T) = _____ A

67. Current through resistor R_1 (I_1) = _____ A

68. Current through resistor R_2 (I_2) = _____ A

69. Current through resistor R_3 (I_3) = _____ A

70. Power converted by resistor R_1 (P_1) = _____ W

71. Power converted by resistor R_2 (P_2) = _____ W

72. Power converted by resistor R_3 (P_3) = _____ W

73. Total power (P_T) = _____ W

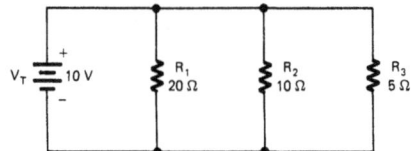

FIGURE 3SE-9

Answers

54. 3.33 Ω	55. 3 A
56. 2 A	57. 1 A
58. 1 Ω	59. 12 A
60. 6 A	61. 3 A
62. 2 A	63. 1 A
64. 144 W	65. 2.857 Ω
66. 3.5 A	67. 0.5 A
68. 1 A	69. 2 A
70. 5 W	71. 10 W
72. 20 W	73. 35 W

Combination Electric Circuits

Combination electric circuits are made up of both series and parallel parts. They are sometimes called *series-parallel circuits*. Almost all electric equipment has combination circuits rather than only series circuits or only parallel circuits. However, it is important to understand series *and* parallel circuits to work with combination circuits.

A series circuit has only one path for current flow from the voltage source. A parallel circuit has two or more paths. In the circuit shown in Fig. 3-23, R_1 is in series with the voltage source, and R_2 and R_3 are in parallel. There are many different types of combination circuits. Some have only one series component and many parallel components. Others have many series components and only a few parallel components. In the circuit shown in Fig. 3-24, R_1 and R_2 are in series with the voltage source. R_3 to R_4 and R_5 to R_6 are in parallel. There are two paths in each parallel part of the

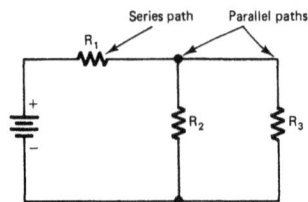

FIGURE 3-23 Simple combination circuit.

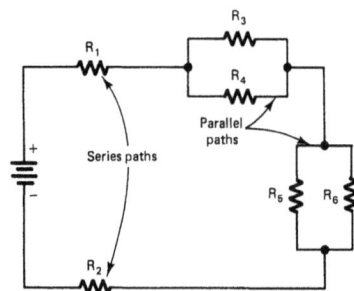

FIGURE 3-24 Combination circuit.

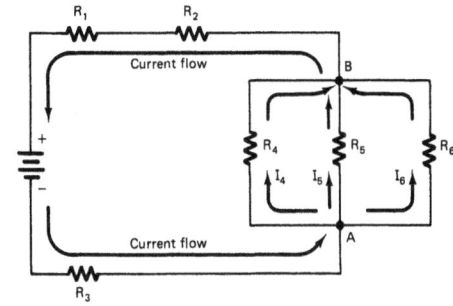

FIGURE 3-25 Current paths in a combination circuit.

circuit. The total current of the circuit flows through each series part of the circuit. In this circuit, the current is the same through resistors R_1 and R_2.

Another combination circuit is shown in Fig. 3-25. R_1, R_2, and R_3 are in series with the voltage source. The total current (I_T) flows through R_1, R_2, and R_3. At point A, the current divides through the parallel paths of R_4, R_5, and R_6. The currents I_4, I_5, and I_6 flow through the parallel paths.

To find the *total resistance* of a combination circuit, the series resistance is added to the parallel resistance. In the circuit shown in Fig. 3-26, R_1 is the only series part of the circuit. R_2 and R_3 are in parallel. First the parallel resistance of R_2 and R_3 is found. Then the series resistance and parallel resistance are added to find the total resistance. Total resistance of this circuit is 4 Ω.

To find the *total current* in the circuit, the method described earlier is used. Current is equal to voltage divided by resistance. The current that flows from the voltage source in this circuit is 2.5 A. The total current also flows through each series part of the circuit. So 2.5 A also flows through R_1. In this circuit, it is easy to find the current flow through R_2 and R_3. Because R_2 and R_3 have equal resistance values, the same current flows through each of them. There is 2.5 A of current flowing to point A of the circuit. This 2.5 A divides into two paths through R_2 and R_3. To find the current through R_2 and R_3, divide the current coming into point A (2.5 A) by the number of paths (2). The current through R_2 and R_3 is 1.25 A.

To find the voltage across R_1, multiply the current through R_1 by the value of R_1. The voltage across R_1 is 2.5 A times 2 Ω, or 5 V. This is the voltage from point C to point B. The voltage across R_2 or R_3 is also across points B and A. The voltage across points B and A is found by means of subtracting the voltage across R_1 from the applied voltage. The voltage across R_2 and R_3 is 10 V – 5 V, or 5 V.

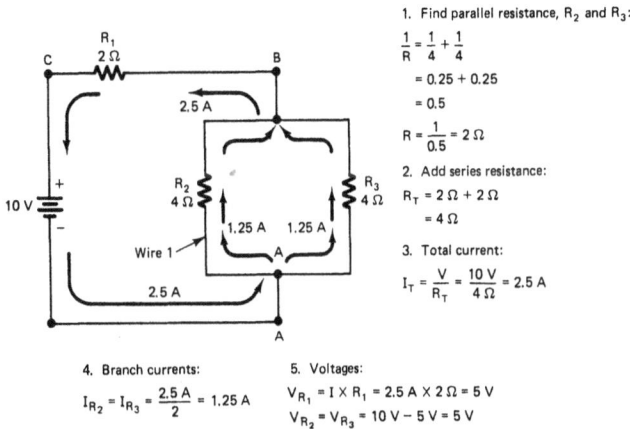

1. Find parallel resistance, R_2 and R_3:

$$\frac{1}{R} = \frac{1}{4} + \frac{1}{4}$$
$$= 0.25 + 0.25$$
$$= 0.5$$
$$R = \frac{1}{0.5} = 2\,\Omega$$

2. Add series resistance:

$$R_T = 2\,\Omega + 2\,\Omega$$
$$= 4\,\Omega$$

3. Total current:

$$I_T = \frac{V}{R_T} = \frac{10\,V}{4\,\Omega} = 2.5\,A$$

4. Branch currents:

$$I_{R_2} = I_{R_3} = \frac{2.5\,A}{2} = 1.25\,A$$

5. Voltages:

$$V_{R_1} = I \times R_1 = 2.5\,A \times 2\,\Omega = 5\,V$$
$$V_{R_2} = V_{R_3} = 10\,V - 5\,V = 5\,V$$

FIGURE 3-26 Example of a combination circuit.

Examples of Combination Circuits

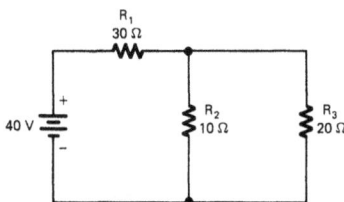

FIGURE 3-27 Example of a combination circuit.

Look at the circuit of Fig. 3-27. The value that should first be calculated is the resistance of R_2 and R_3 in parallel. When this quantity is found, it can be added to the value of the series resistor (R_1) to find the total resistance of the circuit, as follows:

$$R_T = R_1 = R_2 \,||\, R_3$$

(The symbol between R_2 and R_3 means R_2 is in parallel with R_3)

$$= 30\Omega + \frac{10 \times 20}{10 + 20} = 30\Omega + 6.67\Omega = 36.67\Omega$$

When the total resistance (R_T) is found, the total current (I_T) may be found, as follows:

$$I_T = \frac{V_T}{R_T} = \frac{40V}{36.67\Omega} = 1.09A$$

The total current flows through resistor R_1 because it is in series with the voltage source. The voltage drop across resistor R_1 is as follows:

$$V_1 = I_T \times R_1 = 1.09\,A \times 30\,\Omega = 32.7\,V$$

The applied voltage is 40 V, and 32.7 V is dropped across resistor R_1. The remaining voltage is dropped across the two parallel resistors (R_2 and R_3): 40 V – 32.7 V = 7.3 V across R_2 and R_3. The currents through R_2 and R_3 are as follows:

$$I_2 = \frac{V_2}{R_2} = \frac{7.3V}{10\Omega} = 0.73A$$

$$I_3 = \frac{V_3}{R_3} = \frac{7.3V}{20\Omega} = 0.365A$$

Another type of combination circuit is shown in Fig. 3-28. Resistors R_2 and R_3 are in series. When they are combined by means of adding their values, the circuit becomes a two-branch parallel circuit. Total resistance (R_T) is found with the product-over-sum method, as follows:

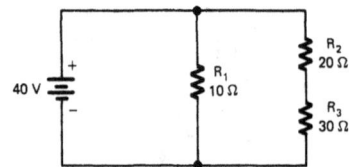

FIGURE 3-28 Example of a combination circuit.

$$R_T = \frac{10 \times 50}{10 + 50} = \frac{500}{60} = 8.33\Omega$$

Total current is found as follows:

$$I_T = \frac{V_T}{R_T} = \frac{40V}{8.33\Omega} = 4.8A$$

The voltage across R_1 is 40 V because it is in parallel with the voltage source. The voltage drops across R_2 and R_3 are as follows:

$$I_2 = \frac{V_T}{R\ of\ branch} = \frac{40V}{50\Omega} = 0.8A$$

$$V_2 = I \times R_2$$

$$= 0.8\,A \times 20\,\Omega$$

$$= 16\,V$$

$$V_3 = I \times R_3$$

$$= 0.8\,A \times 30\,\Omega$$

$$= 24\,V$$

V_2 and V_3 (16 V + 24 V) equal the source voltage.

Combination circuit problems may be solved with the following step-by-step procedure:

1. Combine series and parallel parts to find the total resistance of the circuit.

2. Find the total current that flows through the circuit.

3. Find the voltage across each part of the circuit.

4. Find the current through each resistance of this circuit.

Steps 3 and 4 often must be done in combination with each other, rather than one then the other.

Kirchhoff's Laws

Ohm's law shows the relation of voltage, current, and resistance in electric circuits. *Kirchhoff's voltage law* also is important in solving electric problems. A scientist named Gustav Kirchhoff is given credit for discovering this effect. He found that the sum of voltage drops around any closed-circuit loop must equal the voltage applied to that loop. Figure 3-29a shows a simple series circuit to illustrate this law. This law holds true for any series circuit loop.

Kirchhoff's current law also is important in solving electric problems, especially for parallel circuits. The current law states that at any junction of electric conductors in a circuit, the total amount of current entering the junction must equal the amount of current leaving the junction. Figure 3-29b shows some examples of Kirchhoff's current law.

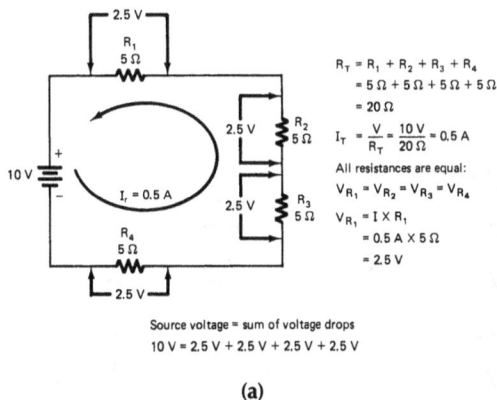

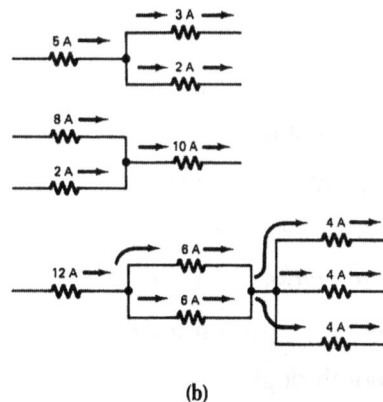

$$R_T = R_1 + R_2 + R_3 + R_4$$
$$= 5\,\Omega + 5\,\Omega + 5\,\Omega + 5\,\Omega$$
$$= 20\,\Omega$$

$$I_T = \frac{V}{R_T} = \frac{10\ V}{20\,\Omega} = 0.5\ A$$

All resistances are equal:

$$V_{R_1} = V_{R_2} = V_{R_3} = V_{R_4}$$

$$V_{R_1} = I \times R_1$$
$$= 0.5\ A \times 5\,\Omega$$
$$= 2.5\ V$$

Source voltage = sum of voltage drops

10 V = 2.5 V + 2.5 V + 2.5 V + 2.5 V

(a)

(b)

FIGURE 3-29 Kirchhoff's laws. (a) Example of voltage law. (b) Examples of current law.

Self-Examination

Solve each of the following combination circuit problems.
Find:

74. Total resistance (R_T) = _____ Ω
75. Total current (I_T) = _____ A
76. Voltage across R_1 (V_1) = _____ V
77. Total power (P_T) = _____ W
78. Current through R_2 (I_2) = _____ A
79. Voltage across R_2 (V_2) = _____ V

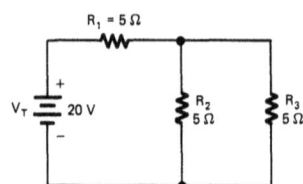

FIGURE 3SE-10

Find:

80. Total resistance (R_T) = _____ Ω
81. Total current (I_T) = _____ A
82. Total power (P_T) = _____ W
83. Voltage across R_1 (V_1) = _____ V
84. Current through R_2 (I_2) = _____ A
85. Current through R_4 (I_4) = _____ A
86. Voltage across R_5 (V_5) = _____ V

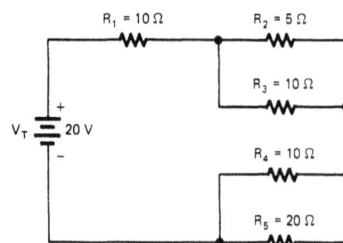

FIGURE 3SE-11

Find:

87. Total resistance (R_T) = _____ Ω
88. Total current (I_T) = _____ A
89. Voltage across resistor R_1 (V_1) = _____ V
90. Voltage across resistor R_4 (V_4) = _____ V
91. Current through resistor R_2 (I_2) = _____ A
92. Current through resistor R_3 (I_3) = _____ A

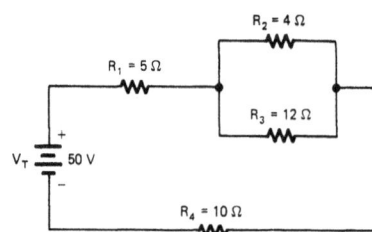

FIGURE 3SE-12

Find:

93. Total resistance (R_T) = _____ Ω
94. Total current (I_T) = _____ A
95. Voltage across resistor R_1 (V_1) = _____ V
96. Voltage across resistor R_2 (V_2) = _____ V
97. Voltage across resistor R_3 (V_3) = _____ V
98. Current through resistor R_3 (I_3) = _____ A
99. Current through resistor R_2 (I_2) = _____ A
100. Total power (P_T) = _____ W
101. Power converted by R_3 (P_3) = _____ W

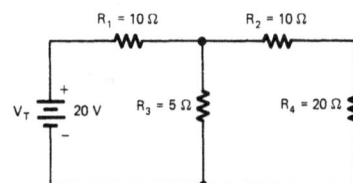

FIGURE 3SE-13

74.	7.5 Ω	75.	2.67 A
76.	13.35 V	77.	53.4 W
78.	1.33 A	79.	6.65 V
80.	20 Ω	81.	1 A
82.	20 W	83.	10 V
84.	0.67 A	85.	0.67 A
86.	6.67 V	87.	18 W
88.	2.78 A	89.	13.9 V
90.	27.8 V	91.	8.3 V
92.	0.69 A	93.	14.28 Ω
94.	1.4 A	95.	14 V
96.	2 V	97.	6 V
98.	1.2 A	99.	0.2 A
100.	28 W	101.	7.2 W

Power in Electric Circuits

In terms of voltage and current, power (P) is equal to voltage (in volts) multiplied by current (in amperes). The formula is: $P = V \times I$. This formula is an easy way to find electric power. For example, a 120 V electric outlet with 4 A of current flowing from it has the following power value

$$P = V \times I \text{ or } P = 120\,V \times 4\,A = 480\,W$$

The unit of electric power is the *watt*. In the example, 480 W of power is converted by the load portion of the circuit. Another way to find power is as follows:

$$P = \frac{V^2}{R}$$

This formula is used when voltage and resistance are known but current is not known. The formula $P = I^2 \times R$ is used when current and resistance are known.

Several formulas are summarized in Fig. 3-30. The quantity in the center of the circle may be found with any of the three formulas in the outer part of the same quadrant in the circle. This circle is handy to use for making electric calculations for voltage, current, resistance, or power.

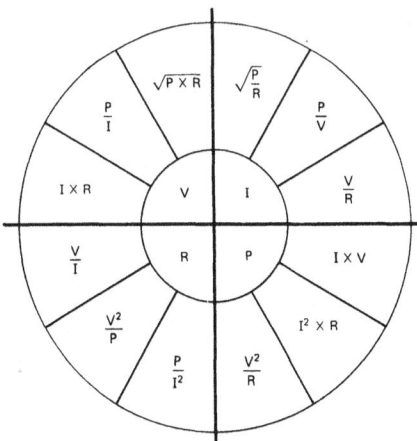

FIGURE 3-30 Formulas for finding voltage, current, resistance, and power.

It is easy to find the amount of power converted by each of the resistors in a series circuit, such as the one shown in Fig. 3-31. In the circuit shown, the amount of power converted by each of the resistors and the total power are found as follows:

1. Power converted by resistor R_1:

$$P_1 = I^2 \times R_1 = 2^2 \times 20\,\Omega = 80\,\text{W}$$

2. Power converted by resistor R_2:

$$P_2 = I^2 \times R_2 = 2^2 \times 30\,\Omega = 120\,\text{W}$$

3. Power converted by resistor R_3:

$$P_3 = I^2 \times R_3 = 2^2 \times 50\,\Omega = 200\,\text{W}$$

4. Power converted by the circuit:

$$P_T = P_1 + P_2 + P_3 = 80\,\text{W} + 120\,\text{W} + 200\,\text{W}$$

$$= 400\,\text{W}$$

or

$$P_T = V_T \times I = 200\,\text{V} \times 2\,\text{A} = 400\,\text{W}$$

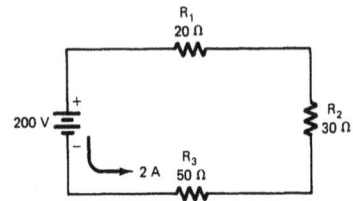

FIGURE 3-31 Finding power values in a series circuit.

When working with electric circuits it is possible to check your results with other formulas.

Power in parallel circuits is found in the same way as for series circuits. In Fig. 3-32, the power converted by each of the resistors and the total power of the parallel circuit is found as follows:

1. Power converted by resistor R_1:

$$P_1 = \frac{V^2}{R_1} = \frac{30^2}{5} = \frac{900}{5} = 180\text{W}$$

2. Power converted by resistor R_2:

$$P_2 = \frac{V^2}{R_2} = \frac{30^2}{10} = \frac{900}{10} = 90\text{W}$$

3. Power converted by resistor R_3:

$$P_3 = \frac{V^2}{R_3} = \frac{30^2}{20} = \frac{900}{20} = 45\text{W}$$

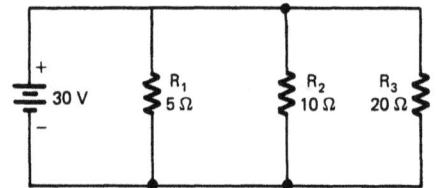

FIGURE 3-32 Finding power values in a parallel circuit.

4. Total power converted by the circuit:

$$P_T = P_1 + P_2 + P_3 = 180\,\text{W} + 90\,\text{W} + 45\,\text{W} = 315\,\text{W}$$

To determine an actual quantity of electric energy, a factor must be used that indicates how long a power value continues. One such unit of electric energy is called a *watt-second* (Ws). It is the product of watts (W) and time (in seconds). The watt-second is a very small quantity of energy. It is more common to measure electric energy in *kilowatt-hours* (kWh). The number of kilowatt-hours of electric energy is used to determine the amount of electric utility bills. A kilowatt-hour is 1000 W per hour, or 3,600,000 Ws.

For example, if an electric heater operates on 120 V and has a resistance of 20 Ω, what is the cost to use the heater for 200 h at a cost of 5¢ per kilowatt-hour?

1. $P = \dfrac{V^2}{R} = \dfrac{120^2}{20\Omega} = \dfrac{14,400}{20\Omega} = 0.72\ \text{kW}$

2. There are 1000 W in a kilowatt (1000 W = 1 kW).

3. Multiply the kilowatts that the heater has used by the hours of use as follows:

   kilowatts × hours = kilowatt-hours (kWh)

   0.72 × 200 h = 144 kWh

4. Multiply the number of kilowatt-hours by the cost.

   Kilowatt-hours × cost = 144 kWh × $0.05 = $7.20

Some simple examples of electric circuits are discussed in this unit. They become easy to understand after practice with each type of circuit. It is important to understand the characteristics of series, parallel, and combination circuits.

Voltage-Divider Circuits

One of the most basic types of electronic circuits is the *voltage divider*. Figure 3-33 shows some types of voltage dividers. The purpose of a voltage-divider circuit is to produce specific values of voltage from one voltage source. The simple series circuit in Figure 3-33a is a voltage divider. Voltage division takes place because of voltage drops across the three resistors. Because each of the three resistors has the same value (1 kΩ), the voltage drop across each one is 3 V. Thus a single voltage source is used to derive three separate voltages of a power supply.

Another method used to accomplish voltage division is the *tapped resistor*. This method relies on the use of a resistor that is wire wound and has a tap onto which a wire is attached. The wire is attached so that a certain amount of total resistance of the device appears from the tap to the outer terminals. For example, if the tap is in the center of a 100 Ω wire-wound resistor, the resistance from the tap to either outer terminal is 50 Ω. Tapped resistors often have two or more taps to obtain several combinations of fixed-value resistance. Figure 3-33b shows a tapped resistor used as a voltage divider. In the example, the voltage outputs are each 3 V, derived from a 6 V source.

A very common method of voltage division is shown in Fig. 3-33c. Potentiometers are used as voltage dividers in volume control circuits of radios and televisions. They may be used to vary voltage from zero to the value of the source voltage. In the example, the voltage output may be varied from 0 to 1.5 V. It is also possible to use a voltage-divider network and a potentiometer to obtain many variable voltage combinations, as discussed later.

Voltage-Divider Design

The design of a voltage-divider circuit is a good application of basic electric theory. Refer to the circuit of Fig. 3-34. Resistors R_1, R_2, and R_3 form a voltage divider to provide the proper voltage to three loads. The loads could be transistors of a 9 V portable radio, for example. The operating voltages and currents of the loads are constant. The values of R_1, R_2, and R_3 are calculated to supply proper voltages to each of the loads. The value of current through R_1 is selected as 10 mA. This value is ordinarily 10% to 20% of the total current flow to the loads (10 mA + 30 mA + 60 mA = 100 mA and 10% of 100 mA = 10 mA). The purpose of R_1 is to provide a closed-loop circuit for the voltage divider.

To calculate the values of R_1, R_2, and R_3, you have to know the voltage across each resistor and the current through each resistor. Start with R_1 at the bottom of the circuit. The current through R_1 is given (10 mA). The voltage across R_1 is 2 V, because the ground is a zero-voltage reference, and 2 V must be supplied to load 1. The value of R_1, as shown in the procedure of Fig. 3-34, must be 200 Ω (2 V ÷ 10 mA).

Resistor R_2 has a voltage of 3 V across it. Point A has a potential of +2 V, and point B has a potential of +5 V for load 2. The *difference in potential,* or voltage drop, is therefore 5 V − 2 V = 3 V. The current through R_2 is 20 mA. A current of 10 mA flows up through R_1, and 10 mA flows to point A from load 1. These two currents (10 mA + 10 mA = 20 mA) combine and flow through R_2. The value of R_2 must be 150 Ω (3 V/20 mA).

Resistor R_3 has a voltage of 4 V across it (9 V − 5 V = 4 V). The current through R_3 is 50 mA because 20 mA flows upward through R_2 and 30 mA flows from load 2 to point B (20 mA + 30 mA = 50 mA). The value of R_3 must be 80 Ω (4 V/50 mA).

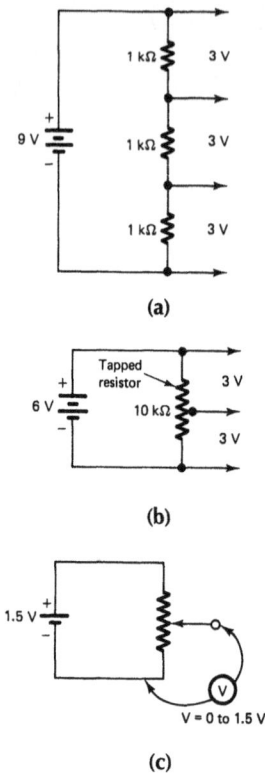

(a)

(b)

(c)

FIGURE 3-33 Voltage divider circuits. (a) Series dc circuit used as a voltage divider. (b) Tapped resistor used as a voltage divider. (c) Potentiometer used as a voltage divider.

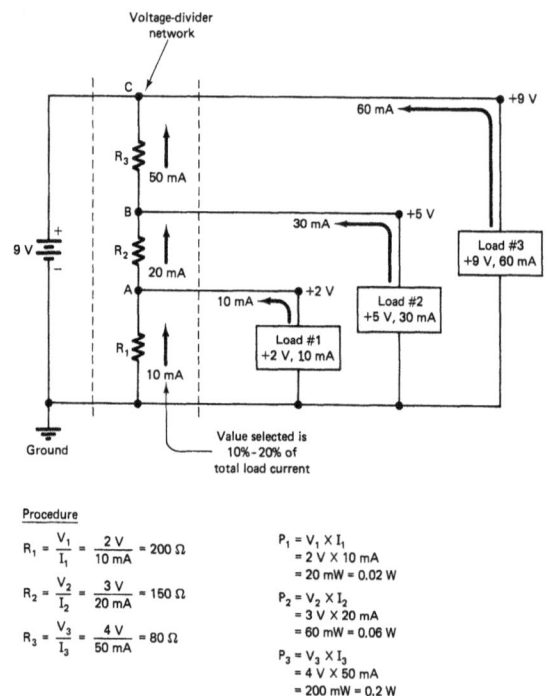

FIGURE 3-34 Voltage divider design.

Procedure

$R_1 = \dfrac{V_1}{I_1} = \dfrac{2\ V}{10\ mA} = 200\ \Omega$

$R_2 = \dfrac{V_2}{I_2} = \dfrac{3\ V}{20\ mA} = 150\ \Omega$

$R_3 = \dfrac{V_3}{I_3} = \dfrac{4\ V}{50\ mA} = 80\ \Omega$

$P_1 = V_1 \times I_1$
$= 2\ V \times 10\ mA$
$= 20\ mW = 0.02\ W$

$P_2 = V_2 \times I_2$
$= 3\ V \times 20\ mA$
$= 60\ mW = 0.06\ W$

$P_3 = V_3 \times I_3$
$= 4\ V \times 50\ mA$
$= 200\ mW = 0.2\ W$

$$V_x = \frac{R_x}{R_T} \times V_T$$

V_x is voltage across a resistance
V_T is total voltage applied
R_x is the resistance where V_x is measured
R_T is the total resistance of the voltage divider network

(a)

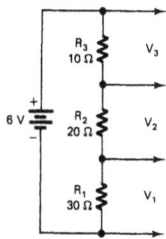

Procedure

$$V_1 = \frac{R_1}{R_T} \times V_T = \frac{30\,\Omega}{60\,\Omega} \times 6\,V = \frac{1}{2} \times 6\,V = 3\,V$$

$$V_2 = \frac{R_2}{R_T} \times V_T = \frac{20\,\Omega}{60\,\Omega} \times 6\,V = \frac{1}{3} \times 6\,V = 2\,V$$

$$V_3 = \frac{R_3}{R_T} \times V_T = \frac{10\,\Omega}{60\,\Omega} \times 6\,V = \frac{1}{6} \times 6\,V = 1\,V$$

(b)

FIGURE 3-35 Voltage division equation and sample problem. (a) Equation. (b) Sample problem.

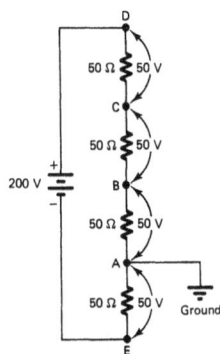

Voltage combinations (taken from ground reference)

$V_{AB} = 50\,V$
$V_{AC} = 100\,V$
$V_{AD} = 150\,V$
$V_{AE} = \boxed{-50\,V}$

FIGURE 3-36 Negative voltage derived from a voltage divider.

With the calculated values of R_1, R_2, and R_3 used as a voltage-divider network, the proper values of voltage are supplied to the three circuit loads. The minimum power rating of each resistor must be considered. Minimum power values of 0.02 W, 0.06 W, and 0.2 W are calculated in Fig. 3-34. A *safety factor* often is used to ensure that power values are large enough. A safety factor is a multiplier used with the minimum power values. For example, if a safety factor of 2 is used, the minimum power values for the circuit are $P_1 = 0.02\,W \times 2 = 0.04\,W$, $P_2 = 0.06\,W \times 2 = 0.12\,W$, $P_3 = 0.2\,W \times 2 = 0.4\,W$.

Voltage-Division Equation

The voltage division equation, often called the *voltage-divider rule,* is convenient to use with voltage-divider circuits. The voltage divider equation and a sample problem are shown in Fig. 3-35. This equation applies to series circuits. The voltage (V_x) across any resistor in a series circuit is equal to the ratio of that resistance (R_x) to total resistance (R_T) multiplied by the source voltage (V_T).

Negative Voltage Derived from a Voltage-Divider Circuit

Voltage-divider circuits often are used as power for electronic equipment. In electronics, reference often is made to *negative* voltage. The concept of negative voltage is made clear in Fig. 3-36. Voltage ordinarily is measured with respect to a ground reference point. The circuit ground is shown at point A. The actual ground reference for electronic circuits often is a metal chassis or ground strip on a printed circuit board (PCB) to simplify testing and troubleshooting. The negative terminal of a meter can be connected directly to the ground point. All voltage readings are taken with respect to the ground reference. Point E in Fig. 3-36 is connected to the negative side of the power source. Point A, where the ground reference is connected, is more positive than point E. Therefore the voltage across points A and E is −50 V.

Voltage Division with a Potentiometer

A sample problem with a pot is shown in Fig. 3-37. A given value of 10 kΩ is used as the pot. The desired variable voltage from the pot center terminal to ground is 5 V to 10 V. The values of R_1 and R_3 are calculated to derive the desired variable voltage from the pot.

The current flow in a voltage-divider network is established by the value of R_2 (10 kΩ) and the range of voltage variation (5 V to 10 V = 5 V variation). The current-flow calculation in the circuit of Fig. 3-37 is shown in the procedure. Because $I = V/R$, the current through R_2 and the other parts of this series circuit is 0.5 mA.

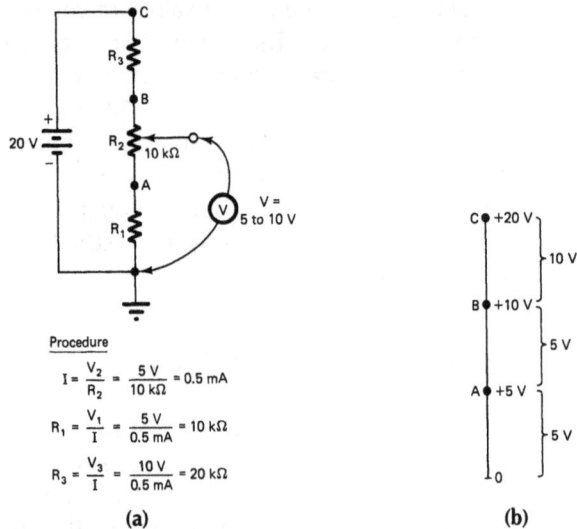

$$I = \frac{V_2}{R_2} = \frac{5\text{ V}}{10\text{ k}\Omega} = 0.5\text{ mA}$$

$$R_1 = \frac{V_1}{I} = \frac{5\text{ V}}{0.5\text{ mA}} = 10\text{ k}\Omega$$

$$R_3 = \frac{V_3}{I} = \frac{10\text{ V}}{0.5\text{ mA}} = 20\text{ k}\Omega$$

(a)   (b)

FIGURE 3-37 Voltage division with a potentiometer. (a) Circuit. (b) Voltage drops across each resistance considered as a scale.

Once the current is found, values of R_1 and R_3 may be found, as shown in the procedure. Figure 3-37b shows an easy method for determining voltage drops. A network of resistances in series can be thought of as a scale. In the example, the voltage at point A is +5 V, and the voltage at point B is +10 V. The difference in potential is 5 V (10 V − 5 V = 5 V). This is similar to reading a scale.

Self-Examination

Solve each of the following voltage-divider problems:

102. Refer to Fig. 3-38 and calculate the values of R_1, R_2, and R_3 needed to form the voltage-divider circuit. Use the voltage divider resistance values of Fig. 3-38.

103. Calculate the minimum power ratings of R_1, R_2, and R_3.

104. Solve for the values of R_1 and R_2 in the voltage-divider circuit of Fig. 3-39.

105. Use the resistance values of Fig. 3-39. Calculate the power ratings of R_1 and R_2 with a safety factor of *three* used.

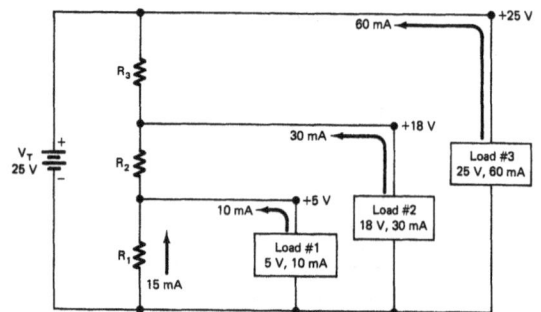

FIGURE 3-38

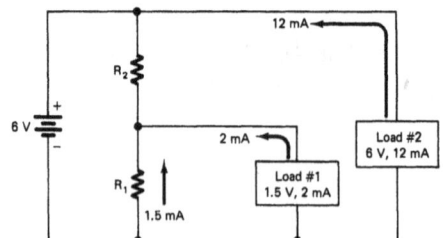

FIGURE 3-39

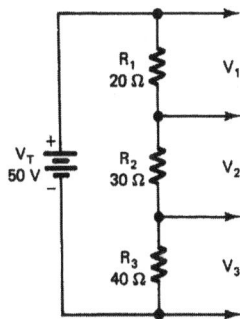

FIGURE 3-40

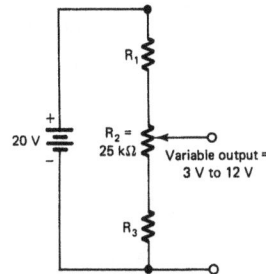

FIGURE 3-41

106. Use the voltage-division equation to find the values of V_1, V_2, and V_3 in Fig. 3-40.

107. A voltage-divider design problem is illustrated in Fig. 3-41. Compute the values of R_1 and R_3 needed to obtain a 3 V to 12 V variable output.

Answers

102. $R_1 = 333.33\ \Omega$, $R_2 = 520\ \Omega$, $R_3 = 127.27\ \Omega$	103. $P_1 = 0.075$ W, $P_2 = 0.325$ W, $P_3 = 0.385$ W
104. $R_1 = 1000\ \Omega$, $R_2 = 1285.7\ \Omega$	105. $P_1 = 0.00675$ W, $P_2 = 0.04725$ W
106. $V_1 = 11.2$ V, $V_2 = 16.6$ V, $V_3 = 22.2$ V	107. $R_1 = 22{,}222.2\ \Omega$, $R_3 = 8{,}333.34\ \Omega$

Maximum Power Transfer

An important consideration in electronic circuit design is maximum power transfer. Maximum power is transferred from a voltage source to a load when the load resistance (R_L) is equal to the internal resistance of the source (R_S). The source resistance limits the amount of power that can be applied to a load. For example, as a flashlight battery ages, its internal resistance increases. This increase in internal resistance causes the battery to supply less power to the lamp load. Thus the light output of the flashlight is reduced.

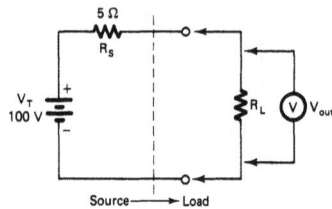

R_L	$I_L = \dfrac{V_T}{R_s + R_L}$	$V_{out} = I_L \times R_L$	$P_{out} = I_L \times V_{out}$
0	$\dfrac{100\text{ V}}{5\,\Omega} = 20$ A	20 A $\times$ 0 Ω = 0 V	20 A $\times$ 0 V = 0 W
2.5 Ω	$\dfrac{100\text{ V}}{7.5\,\Omega} = 13.3$ A	13.3 A $\times$ 2.5 Ω = 33.3 V	13.3 A $\times$ 33.3 V = 449 W
5 Ω	$\dfrac{100\text{ V}}{10\,\Omega} = 10$ A	10 A $\times$ 5 Ω = 50 V	10 A $\times$ 50 V = 500 W
7.5 Ω	$\dfrac{100\text{ V}}{12.5\,\Omega} = 8$ A	8 A $\times$ 7.5 Ω = 60 V	8 A $\times$ 60 V = 480 W
10 Ω	$\dfrac{100\text{ V}}{15\,\Omega} = 6.7$ A	6.7 A $\times$ 10 Ω = 67 V	6.7 A $\times$ 67 V = 444 W

Graph showing maximum power transfer

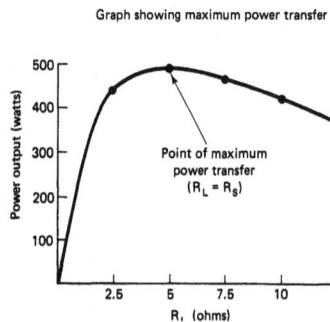

FIGURE 3-42 Problem that shows maximum power transfer.

Figure 3-42 shows an example of maximum power transfer. The source is a 100 V battery with an internal resistance of 5 Ω. The values of I_L, V_{out}, and power output (P_{out}) are calculated as follows:

$$I_L = \frac{V_T}{R_S + R_L}; \quad V_{out} = I_L \times R_L; \quad P_{out} = I_L \times V_{out}$$

The graph in Fig. 3-42 shows that maximum power is transferred from the source to the load when $R_L = R_S$. This is an important circuit design consideration for power sources, amplifier circuits, microphones, or practically any type of electronic circuit.

Kirchhoff's Voltage Law Problems

Kirchhoff's voltage law is illustrated in two different ways in Fig. 3-43. The voltage law may be stated in the two following ways: (1) The sum of voltage drops in a closed-looped circuit is equal to the source voltage, or (2) the algebraic sum of the voltage sources and voltage drops in a closed-loop path is equal to zero. The first method deals with the voltage drops in a closed-loop, or series, path. The sum of the voltage drops across the

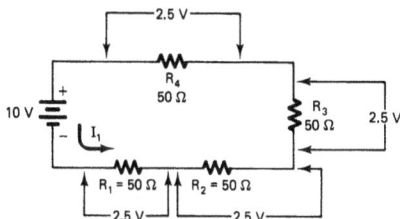

(a) Voltage drop procedure:

$10\text{ V} = 2.5\text{ V} + 2.5\text{ V} + 2.5\text{ V} + 2.5\text{ V}$

(b) Algebraic procedure:

$$10\text{ V} - 50I_1 - 50I_1 - 50I_1 - 50I_1 = 0$$
$$10\text{ V} - 200I_1 = 0$$
$$I_1 = \frac{10\text{ V}}{200} = 0.05\text{ A}$$

FIGURE 3-43 Kirchhoff's voltage law.

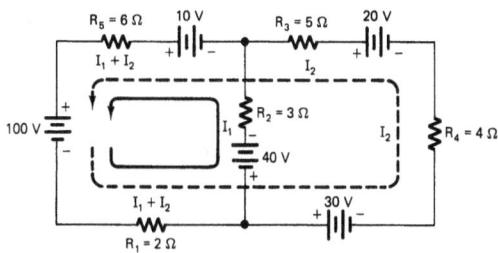

(a) Circuit

(b) Procedure

1. Assign a direction of current flow (− to +) through each path. Start with the largest voltage source.

2. Develop an equation for loop 1 (solid line):

$100\text{ V} - 2(I_1 + I_2) + 40\text{ V} - 3I_1 - 10\text{ V} - 6(I_1 + I_2) = 0$

$100\text{ V} - 2I_1 - 2I_2 + 40\text{ V} - 3I_1 - 10\text{ V} - 6I_1 - 6I_2 = 0$ (Multiply $I_1 + I_2$ terms)

Equation for loop 1 $\longrightarrow$ $\boxed{130\text{ V} - 11I_1 - 8I_2 = 0}$ (Combine like terms)

3. Develop an equation for loop 2 (dashed line):

$100\text{ V} - 2(I_1 + I_2) + 30\text{ V} - 4I_2 - 20\text{ V} - 5I_2 - 10\text{ V} - 6(I_1 + I_2) = 0$

$100\text{ V} - 2I_1 - 2I_2 + 30\text{ V} - 4I_2 - 20\text{ V} - 5I_2 - 10\text{ V} - 6I_1 - 6I_2 = 0$

(Multiply $I_1 + I_2$ terms)

Equation for loop 2 $\longrightarrow$ $\boxed{100\text{ V} - 8I_1 - 17I_2 = 0}$ (Combine like terms

4. Place the two equations together and eliminate either I_1 or I_2 to solve for one unknown term:

(× 8) $130\text{ V} - 11I_1 - 8I_2 = 0$ ⎫ Multiply top equation by 8 and

(× 11) $100\text{ V} - 8I_1 - 17I_2 = 0$ ⎬ bottom equation by 11 to eliminate I_1

5. Rewrite the equations after they have been multiplied:

$$1040\text{ V} - 88I_1 - 64I_2 = 0$$
$$(-)\ 1100\text{ V} \overset{(+)}{-} 88I_1 \overset{(+)}{-} 187I_2 = 0 \longleftarrow$$ 6. Subtract bottom equation
$$-\quad 60\text{ V} \quad\uparrow\quad +123I_2 = 0 \qquad\text{from top equation}$$

(These cancel)

7. Solve for I_2:

$$-60\text{ V} + 123I_2 = 0$$
$$+ 123I_2 = 60\text{ V}$$
$$I_2 = \frac{60\text{ V}}{123} = 0.488\text{ A}$$

8. Substitute value of I_2 in one of the original equations to solve for I_1:

$$100\text{ V} - 8I_1 - 17I_2 = 0$$
$$100\text{ V} - 8I_1 - 17(0.488) = 0$$
$$100\text{ V} - 8I_1 - 8.3\text{V} = 0$$
$$91.7\text{ V} - 8I_1 = 0$$
$$- 8I_1 = -91.7\text{ V}$$
$$I_1 = \frac{-91.7\text{ V}}{8}$$
$$I_1 = 11.46\text{ A}$$

9. Add I_1 and I_2 to get I_T:

$$I_1 + I_2 = I_T$$
$$11.46\text{ A} + 0.488\text{ A} = 11.948\text{ A}$$

FIGURE 3-44 Example of voltage law.

components is equal to the source voltage. In the example in Fig. 3-43, voltage drops are written as $R \times I$, such as $50\ \Omega \times I_1 = 50I_1$. The current in a loop is given an algebraic value (I_1). Remember that any voltage drop is equal to $I \times R$. The algebraic procedure of Kirchhoff's voltage law involves setting up a simple equation for a circuit loop. Values of current flow in circuits may be found with this procedure. In a circuit that has only one voltage source, it is easier to use Ohm's law for series circuits to find current flow.

Kirchhoff's Voltage Method for Problem Solving

The advantage of the algebraic method for problem solving is that currents in multiple-source circuits can be easily calculated. Ohm's law cannot be used to find the current flow through each of the paths shown in the circuit of Fig. 3-44a. The algebraic procedure derived from Kirchhoff's voltage law allows calculation of current in a circuit with more than one voltage source.

The method used in Fig. 3-44 may be used for multiple voltage-source problems that have two current loops. The first step in this method is to assign directions of current flow (from − to +) in the circuit. When there are several sources, start with the largest voltage source. To avoid confusion, the current paths should be marked so that they appear different. The example uses a solid line for path 1 (I_1) and a dashed line for path 2 (I_2). When both paths pass through a resistance, the current is called $I_1 + I_2$.

An equation is developed for each of the circuit loops on the basis of Kirchhoff's voltage law. Each current is followed from the largest voltage source in the direction of the current arrow. Voltage sources must be given the proper sign when an equation is set up. When the current direction is to + through the source, a negative (−) sign is used in the equation. A positive (+) sign is used when the direction of the current arrow is from + to − through the source. The equation for each loop is developed with simple algebraic procedures as shown. Practice in using this method makes it a convenient way to calculate current flow in a circuit with two current loops and two or more voltage sources. Another example is shown in Fig. 3-45.

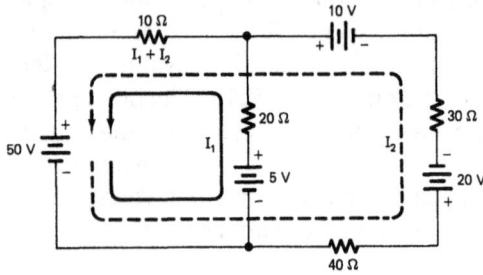

FIGURE 3-45 Kirchhoff's voltage law example problem.

Superposition Method

An alternative method for finding current flow in circuits with two or more voltage sources is called the *superposition method*. This is a nonalgebraic method that involves some rather lengthy but simple calculations. Multiple-voltage-source circuits may be broken down into as many individual circuits as there are voltage sources. For example, a circuit with two voltage sources is reduced to two individual circuits. Each voltage source is considered separately; other voltage sources are short-circuited for making current calculations. In this way the contribution of each voltage source to the current flow in the circuit may be determined. For a two-source circuit, one of the individual circuits is superimposed onto the other with the procedure outlined later.

Figure 3-46 shows a circuit with two voltage sources. The procedure for finding current flow through each component in the circuit is as follows:

1. Short-circuit one power source and use basic Ohm's law procedure to find current flow through each component.

1. Equation for loop 1:

$$50\text{ V} - 5\text{ V} - 20I_1 - 10(I_1 + I_2) = 0$$
$$45\text{ V} - 20I_1 - 10I_1 - 10I_2 = 0$$
$$\boxed{45\text{ V} - 30I_1 - 10I_2 = 0}$$

2. Equation for loop 2:

$$50\text{ V} - 40I_2 + 20\text{ V} - 30I_2 - 10\text{ V} - 10(I_1 + I_2) = 0$$
$$60\text{ V} - 40I_2 - 30I_2 - 10I_1 - 10I_2 = 0$$
$$\boxed{60\text{ V} - 80I_2 - 10I_1 = 0}$$

3. Combine equations and cancel one unknown term:

$$
\begin{aligned}
&\ \ 45\text{ V} - 30I_1 - 10I_2 = 0\\
&(\times 3)\ \ 60\text{ V} - 10I_1 - 80I_2 = 0 \quad \longleftarrow \text{Multiply this equation by 3}\\[4pt]
\hline
&\ \ 45\text{ V} \overset{(+)}{-} 30I_1 \overset{(+)}{-} 10I_2 = 0\\
&(-)\ 180\text{ V} \overset{(+)}{\pm} 30I_1 \overset{(+)}{\pm} 240I_2 = 0 \quad \longleftarrow \begin{array}{l}\text{Subtract bottom equation}\\ \text{from top equation}\end{array}\\[4pt]
\hline
&-135\text{ V} +230I_2 = 0
\end{aligned}
$$

$$230I_2 = 135\text{ V}$$
$$I_2 = \frac{135\text{ V}}{230} = 0.587\text{ A}$$

4. Substitute value of I_2 in one equation to solve for I_1:

$$60\text{ V} - 10I_1 - 80I_2 = 0$$
$$60\text{ V} - 10I_1 - 80(0.587) = 0$$
$$60\text{ V} - 10I_1 - 47\text{ V} = 0$$
$$13\text{ V} - 10I_1 = 0$$
$$-10I_1 = -13\text{ V}$$
$$I_1 = \frac{-13\text{ V}}{10} = 1.3\text{ A}$$

5. Add $I_1 + I_2$ to get I_T:

$$I_1 + I_2 = I_T$$
$$1.3\text{ A} + 0.587\text{ A} = 1.887\text{ A}$$

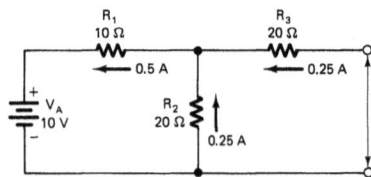

(a)

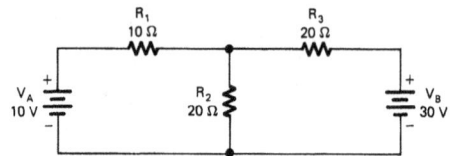

1. $R_T = 10\ \Omega + 20\ \Omega \parallel 20\ \Omega = 20\ \Omega$

2. $I_T = I_1 = \dfrac{V_A}{R_T} = \dfrac{10\text{ V}}{20\ \Omega} = 0.5\text{ A}$

3. $V_1 = I_1 \times R_1$
   $= 0.5\text{ A} \times 10\ \Omega$
   $= 5\text{ V}$

4. $V_2 = V_3 = V_T - V_1$
   $= 10\text{ V} - 5\text{ V}$
   $= 5\text{ V}$

5. $I_2 = \dfrac{V_2}{R_2} = \dfrac{5\text{ V}}{20\ \Omega} = 0.25\text{ A}$

6. $I_3 = \dfrac{V_3}{R_3} = \dfrac{5\text{ V}}{20\ \Omega} = 0.25\text{ A}$

(b)

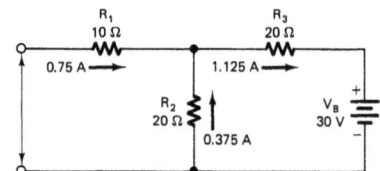

1. $R_T = 20\ \Omega + 20\ \Omega \parallel 10\ \Omega = 26.67\ \Omega$

2. $I_T = I_3 = \dfrac{V_B}{R_T} = \dfrac{30\text{ V}}{26.67\ \Omega} = 1.125\text{ A}$

3. $V_3 = I_3 \times R_3$
   $= 1.125\text{ A} \times 20\ \Omega$
   $= 22.5\text{ V}$

4. $V_2 = V_1 = V_T - V_3$
   $= 30\text{ V} - 22.5\text{ V}$
   $= 7.5\text{ V}$

5. $I_2 = \dfrac{V_2}{R_2} = \dfrac{7.5\text{ V}}{20\ \Omega} = 0.375\text{ A}$

6. $I_1 = \dfrac{V_1}{V_2} = \dfrac{7.5\text{ V}}{10\ \Omega} = 0.75\text{ A}$

(c)

FIGURE 3-46 The superposition method. (a) Original circuit. (b) Circuit with 30 V source shorted. (c) Circuit with 10 V source shorted.

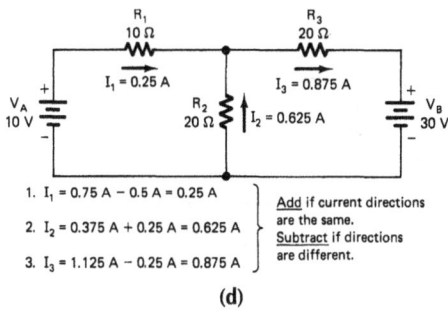

1. $I_1 = 0.75\ A - 0.5\ A = 0.25\ A$

2. $I_2 = 0.375\ A + 0.25\ A = 0.625\ A$ } Add if current directions are the same.
   Subtract if directions are different.

3. $I_3 = 1.125\ A - 0.25\ A = 0.875\ A$

(d)

FIGURE 3-46 (d) Original circuit with currents recorded.

2. Record the amount of current and the direction of flow through each component of this circuit.

3. Short-circuit the other power source and use basic Ohm's law procedures to find current flow through each component.

4. Record the amount of current and the direction of flow through each component of this circuit.

5. Find the current flow through each component by looking at the direction of flow through each circuit. If the directions through components of both circuits are the same, add the values. If the directions of current flow are opposite, subtract the values.

6. Record the amount and direction of current flow on the original circuit. The current flows in the direction of the largest flow in an individual circuit.

The superposition method can be used for circuits with more than two sources. A four-source circuit, for example, would require four individual circuits superimposed to find resultant current values. Current flows in the same directions are added and those in opposite directions are subtracted. Direction of current flow is in the direction of the largest sum of currents through the path.

Equivalent Circuits

The previous sections deal with relatively simple circuit applications. Simplification of more complex circuits may be accomplished by means of applying equivalent circuit methods. Several equivalent circuit methods, sometimes called *complex circuit theorems*, or *network theorems*, may be used to simplify complex circuits. This unit deals primarily with the Thevinin and Norton equivalent circuit applications for solving complex electric-circuit problems.

Thevinin Equivalent Circuit Method

The Thevinin equivalent circuit method is used to simplify electric circuits. A French engineer, M. L. Thevinin, developed this method. Using this method allows a complex circuit to be reduced to one equivalent voltage source and series resistance for purposes of calculation or lab experimentation. It is a very practical method used to calculate load currents and load voltages for any value of load resistance. Working with varying values of load resistance is greatly simplified with the Thevinin equivalent-circuit method.

The Thevinin equivalent circuit is shown in Fig. 3-47. It is called an equivalent circuit because it is equivalent to a more

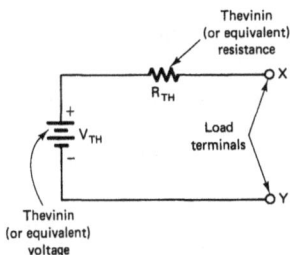

FIGURE 3-47 Thevinin equivalent circuit.

complex circuit (as shown by a load connected to the circuit). Remember that circuits have a source and a load. A complex circuit is reduced to one with a single voltage source (V_{TH}) and a series resistance (R_{TH}). These values are called the *equivalent voltage* and *equivalent resistance*. The load is connected to the load terminals of the circuit, which are labeled points X and Y.

The procedure for simplifying circuits using the Thevinin method is explained next. Examples are shown in Figures 3-48 through 3-51.

Single-Source Problem

The Thevinin equivalent-circuit method may be used for simplifying circuits that have one voltage source. Figure 3-48 shows a circuit with one voltage source and the calculations used to obtain an equivalent circuit. The procedure for finding V_{TH} and R_{TH} is as follows:

1. Find V_{TH}.

   a. Remove the load from the circuit, leaving terminals X and Y open.

   b. Use basic Ohm's law procedures to find the voltage across the load (X and Y) terminals. The voltage across the load is the equivalent voltage (V_{TH}). When resistances are in series with the load, they are disregarded for finding V_{TH}. V_{TH} is an open-circuit voltage; therefore maximum voltage in the current loop that contains the load appears across the load terminals (X and Y).

2. Find R_{TH}.

   a. Replace the source with a short circuit.

   b. Remove the load from the circuit, leaving terminals X and Y open.

   c. "Look into" the circuit from the load terminals to determine the circuit configuration as shown by a load connected to the load terminals. Examples of determining circuit configurations as shown by the load are shown in Fig. 3-49.

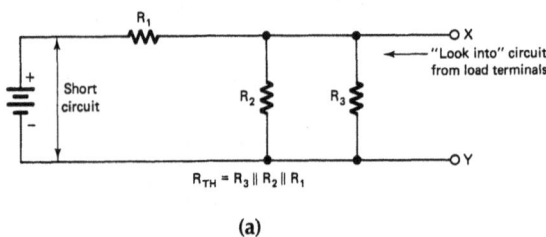

(a)

1. Find V_{TH}:

   a. $R_T = R_1 + R_2 \| R_3$
   $= 5\,\Omega + 10\,\Omega \| 10\,\Omega$
   $= 10\,\Omega$

   b. $I_T = \dfrac{V_T}{R_T} = \dfrac{30\,V}{10\,\Omega} = 3\,A$

   c. $V_{TH} = I_T \times R_2 \| R_3$
   $= 3\,A \times 5\,\Omega$
   $= 15\,V$

2. Find R_{TH}:

   $R_{TH} = R_3 \| R_2 \| R_1$
   $= 10\,\Omega \| 10\,\Omega \| 5\,\Omega$
   $= 2.5\,\Omega$

(b)

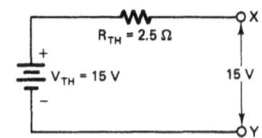

(c)

FIGURE 3-48 Use of the Thevinin method for a one-source circuit. (a) Original circuit. (b) Problem-solving procedure. (c) Thevinin equivalent circuit.

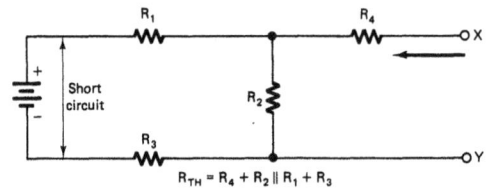

$R_{TH} = R_3 \| R_2 \| R_1$

(a)

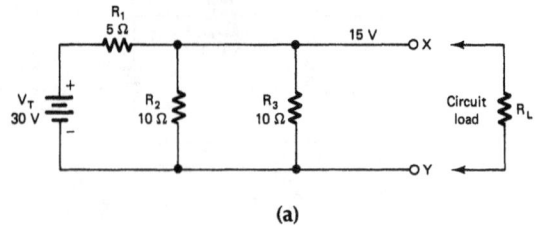

$R_{TH} = R_4 + R_2 \| R_1 + R_3$

(b)

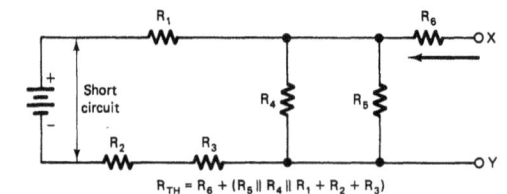

$R_{TH} = R_6 + (R_5 \| R_4 \| R_1 + R_2 + R_3)$

(c)

FIGURE 3-49 Determining circuit configuration for finding R_{TH}.

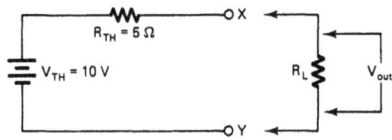

R_L	$I_L = \dfrac{V_{TH}}{R_{TH} + R_L}$	$V_{out} = I_L \times R_L$
5 Ω	$\dfrac{10\,V}{10\,\Omega} = 1\,A$	$1\,A \times 5\,\Omega = 5\,V$
10 Ω	$\dfrac{10\,V}{15\,\Omega} = 0.667\,A$	$0.667\,A \times 10\,\Omega = 6.67\,V$
15 Ω	$\dfrac{10\,V}{20\,\Omega} = 0.5\,A$	$0.5\,A \times 15\,\Omega = 7.5\,V$
20 Ω	$\dfrac{10\,V}{25\,\Omega} = 0.4\,A$	$0.4\,A \times 20\,\Omega = 8\,V$

FIGURE 3-50 Calculating load current and voltage output.

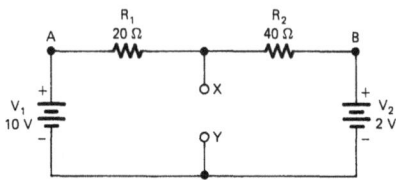

1. Find R_{TH}:

$R_{TH} = R_1 \parallel R_2$
$= 20\,\Omega \parallel 40\,\Omega$
$= 13.33\,\Omega$

2. Find V_{TH}:

Potential at point A = +10 V
Potential at point B = +2 V
Difference in potential = 10 V − 2 V = 8 V
Current flow = $\dfrac{8\,V}{60\,\Omega} = 0.133\,A$

$V_1 = I \times R_1 = 0.133\,A \times 20\,\Omega = 2.67\,V$
$V_{TH} = 10\,V − V_1 = 10\,V − 2.67\,V = 7.33\,V$
or
$V_2 = I \times R_2 = 0.133\,A \times 40\,\Omega = 5.33\,V$
$V_{TH} = 2\,V + V_2 = 2\,V + 5.33\,V = 7.33\,V$

(a)

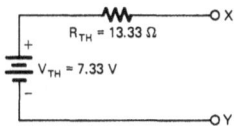

(b)

FIGURE 3-51 Two-source Thevinin equivalent circuit. (a) Problem-solving procedures. (b) Thevinin equivalent circuit.

After the equivalent circuit has been developed, it is simple to calculate values of load current (I_L) and voltage output across a load resistance (V_{out}). Figure 3-50 shows the calculations of several values of I_L and V_{out} with a Thevinin equivalent circuit.

Two-Source Problem

The Thevinin equivalent circuit is easy to apply to a circuit that has two voltage sources. Consider the circuit of Fig. 3-51, which has 10 V and 2 V sources. The load terminals (X and Y) are in the center of the diagram. To find the R_{TH}, look from the load terminals into the circuit. R_1 and R_2 are in parallel as shown by the load terminals.

The Thevinin equivalent voltage (V_{TH}) is found by looking at the difference in potential across the circuit resistances. The same procedure as single-source circuits is used, but the potential at point X must be found. The potential at point X in the example is V_{TH}. The difference in potential across points A and B is 8 V (10 V − 2 V). The 8 V value is used to find the current that would flow through R_1 and R_2. Once the current is calculated, the voltage across either resistor may be found. The voltage across R_1 may be subtracted from the potential at point A to find V_{TH}. The voltage across R_2 may be added to the potential at point B to determine V_{TH}.

If the polarity of V_2 is reversed, the difference in potential across R_1 and R_2 would become 12 V [(10 V − (−2 V)]. This would change the value of V_{TH}.

Thevinin equivalent circuits are used to reduce more complex circuits into a circuit that has one equivalent voltage source and one series resistance. They are very helpful in simplifying the procedure for calculating load current and voltage output of circuits that have several values of load resistance.

Norton Method

Another method of simplifying circuits is the Norton equivalent-circuit method. The Norton equivalent circuit is shown in Fig. 3-52. The Norton current (I_N) is the maximum current that will flow from the source. It is calculated when $R_L = 0\,\Omega$. The Norton resistance (R_N) is calculated in the same way as Thevinin resistance (R_{TH}). The Norton method allows reduction of a circuit to a constant current source (I_N) and an equivalent resistance (R_N) in parallel. Figure 3-53a

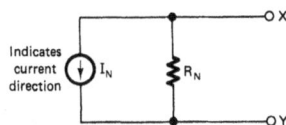

FIGURE 3-52 Norton equivalent circuit.

shows a sample procedure for applying the Norton method. The procedure for developing a Norton equivalent circuit is as follows:

1. Find I_N.

    a. Short-circuit the load (X and Y) terminals.

    b. Calculate the current that will flow from the source when the load resistance is equal to 0 Ω (short circuit). This is the Norton current (I_N).

    c. Label the direction of current flow from the source with an arrow on the equivalent circuit diagram.

2. Find R_N.

    a. Use the same procedure as outlined for finding R_{TH}.

    b. Label the value of R_N on the equivalent circuit diagram.

The load current (I_L) that will flow from a circuit can be calculated easily by means of applying the Norton equivalent circuit. The formula used to calculate load current values from the equivalent values of R_N and I_N is as follows:

$$I_L = \frac{I_N \times R_N}{R_N + R_L}$$

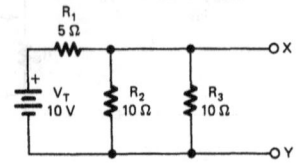

1. Find I_N:
  $R_L = 0\,\Omega$, so R_2 and R_3 are eliminated.
$$I_N = \frac{V_T}{R} = \frac{10\text{ V}}{5\,\Omega} = 2\text{ A}$$

2. Find R_N:
  $R_N = R_3 \| R_2 \| R_1$
    $= 10\,\Omega \| 10\,\Omega \| 5\,\Omega$
    $= 2.5\,\Omega$

(a)

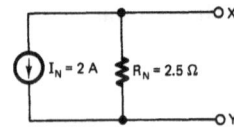

(b)

FIGURE 3-53 Norton equivalent-circuit procedure. (a) Problem-solving procedure. (b) Norton equivalent circuit.

Bridge-Circuit Simplification

A bridge circuit is shown in Fig. 3-54. Bridge circuits are used for several applications, particularly in electric measurement. A bridge circuit may be designed to measure electric component values by means of comparing an unknown value with a known or standard value. Other applications of bridge circuits include rectification circuits, which convert ac into dc.

Bridge circuits are difficult to analyze with Ohm's law techniques. The easiest method to use in developing a Thevinin equivalent circuit is to simplify the analysis of this type of circuit. Use the following procedure to calculate the value of Thevinin equivalent voltage (V_{TH}) and equivalent resistance (R_{TH}) for the bridge circuit shown in Fig. 3-55.

1. Find R_{TH}.

    a. Remove the load resistance from the circuit.

    b. Look into the circuit from the load (X and Y) terminals to determine R_{TH}. The power supply should be replaced by a short circuit.

    c. The circuit arrangement of the four resistors is shown in Fig. 3-55.

    d. Calculate the R_{TH} of this arrangement and label its value on the equivalent-circuit diagram.

2. Find V_{TH}.

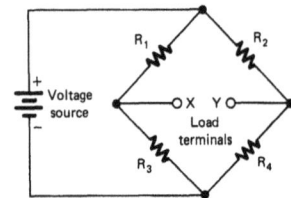

FIGURE 3-54 Bridge circuit.

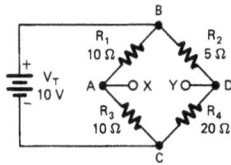

Problem-solving procedure

1. Find R_{TH}:

   a. Resistor configuration looking from load terminals with voltage source circuited is:

   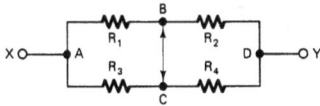

   b. This configuration is the same as:

   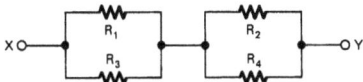

   c. Therefore, $R_{TH} = R_1 \| R_3 + R_2 \| R_4$
      $= 10\,\Omega \| 10\,\Omega + 5\,\Omega \| 20\,\Omega$
      $= 5\,\Omega + 4\,\Omega = 9\,\Omega$

2. Find V_{TH}:

   a. Disregard $R_2 - R_4$ path to find $I_1 = I_3$.
   $$I_1 = I_3 = \frac{V_T}{R_1 + R_3} = \frac{10\text{ V}}{20\,\Omega} = 0.5\text{ A}$$

   b. Disregard $R_1 - R_3$ path to find $I_2 = I_4$:
   $$I_2 = I_4 = \frac{V_T}{R_2 + R_4} = \frac{10\text{ V}}{25\,\Omega} = 0.4\text{ A}$$

   c. Calculate V_3:
   $V_3 = I_3 \times R_3$
   $= 0.5\text{ A} \times 10\,\Omega$
   $= 5\text{ V}$

   b. Calculate V_4:
   $V_4 = I_4 \times R_4$
   $= 0.4\text{ A} \times 20\,\Omega$
   $= 8\text{ V}$

   e. Subtract V_3 from V_4 to find V_{TH}:
   $V_{TH} = V_4 - V_3$
   $= 8\text{ V} - 5\text{ V}$
   $= 3\text{ V}$

3. Complete the Thevinin equivalent circuit diagram:

   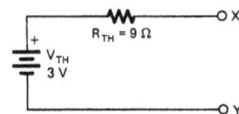

FIGURE 3-55 Simplification of a bridge circuit.

a. Disregard R_2 and R_4 and calculate the current that would flow through R_1 and R_3 if R_2 and R_4 were disconnected from the circuit.

b. Disregard R_1 and R_3 and calculate the current that would flow through R_2 and R_4 if R_1 and R_3 were disconnected from the circuit.

c. Calculate the voltage drop across R_3 ($V_3 = I_3 \times R_3$) with the current determined in step a.

d. Calculate the voltage drop across R_4 caused by the current determined in step b ($V_4 = I_4 \times R_4$).

e. Subtract V_3 from V_4. This is the difference in potential (voltage drop) across points X and Y. This value is the V_{TH} of the circuit, which should be labeled on the equivalent-circuit diagram.

The use of a Thevinin equivalent circuit greatly simplifies calculation of the load current and voltage output of bridge circuits. In Fig. 3-56 addition of a load resistance to a bridge circuit produces a complex circuit configuration. Simplification of a bridge circuit by means of the Thevinin method provides an easy way to analyze bridge circuits.

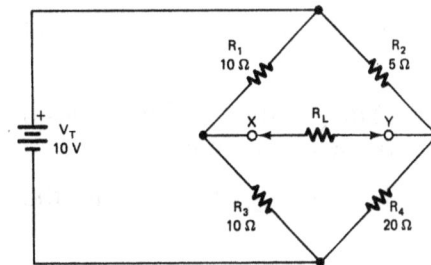

Original bridge circuit

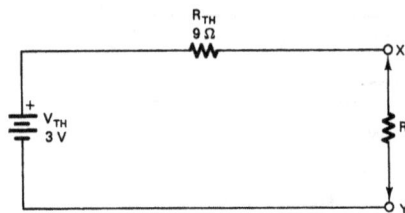

Thevinin equivalent circuit

Load current and output voltage calculations:

R_L	$I_L = \dfrac{V_{TH}}{R_{TH} + R_L}$	$V_{out} = I_L \times R_L$
$3\,\Omega$	$\dfrac{3\text{ V}}{12\,\Omega} = 0.25\text{ A}$	$0.25\text{ A} \times 3\,\Omega = 0.75\text{ V}$
$6\,\Omega$	$\dfrac{3\text{ V}}{15\,\Omega} = 0.2\text{ A}$	$0.2\text{ A} \times 6\,\Omega = 1.2\text{ V}$
$9\,\Omega$	$\dfrac{3\text{ V}}{18\,\Omega} = 0.16\text{ A}$	$0.16\text{ A} \times 9\,\Omega = 1.44\text{ V}$
$12\,\Omega$	$\dfrac{3\text{ V}}{21\,\Omega} = 0.143\text{ A}$	$0.143\text{ A} \times 12\,\Omega = 1.71\text{ V}$

FIGURE 3-56 Calculating load current and voltage output of a bridge circuit.

Self-Examination

Solve the following maximum power-transfer problems.

108. Find the values of load current (I_L), voltage output (V_{out}), and power output (P_{out}) for the circuit in Fig. 3-57 using load resistance values of 0, 1, 2, 3, 4, 5, 6, 7, and 8 Ω.

109. Draw a power-transfer curve using the values obtained for Fig. 3-57. Plot power output (W) on the vertical axis and load resistance (R_L) on the horizontal axis.

110. Find the values of V_{out} and P_{out} for the circuit in Fig. 3-57 if the value of input voltage is changed to 20 V, and R_S is changed to 3 Ω. Use the same R_L values as in the preceding problem.

Solve each of the following problems by applying Kirchhoff's current law to the circuit shown in Fig. 3-58.

111. I_1 in Fig. 3-58a = _____ A

112. I_2 in Fig. 3-58b = _____ A

113. I_3 in Fig. 3-58c = _____ A

114. I_4 in Fig. 3-58d = _____ A

Solve each of the following problems by applying Kirchhoff's voltage law.

115. Use the algebraic method to find values of I_1, I_2, and $I_1 + I_2$ for the circuit in Fig. 3-59.

116. Solve for values of I_1, I_2, and $I_1 + I_2$ for the circuit in Fig. 3-60.

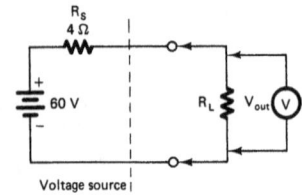

FIGURE 3-57

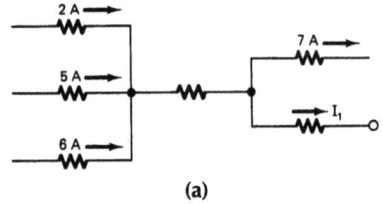

(a)

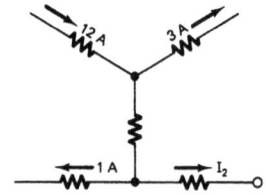

(b)

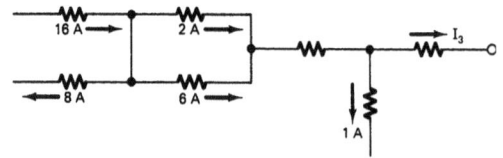

(c)

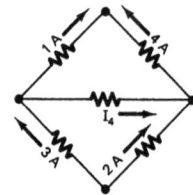

(d)

FIGURE 3-58

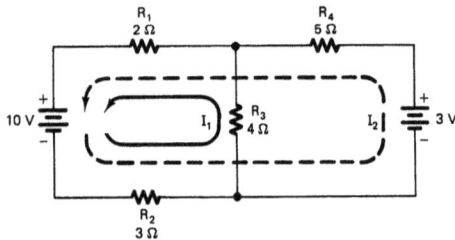

FIGURE 3-59

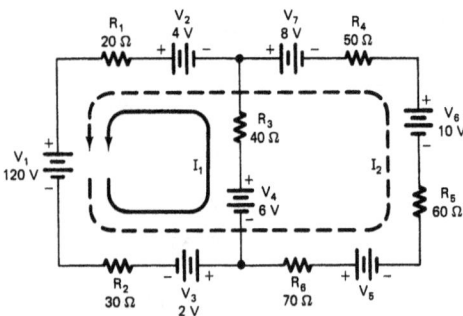

FIGURE 3-60

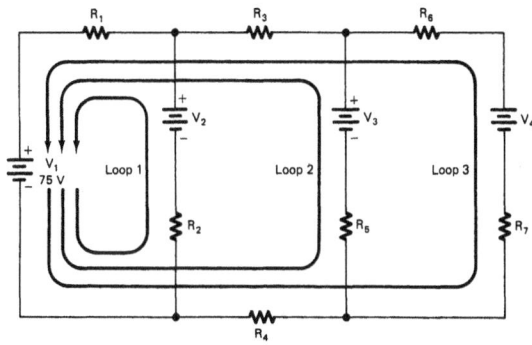

FIGURE 3-61

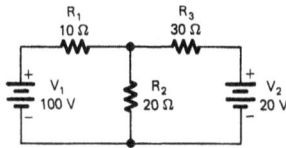

FIGURE 3-62

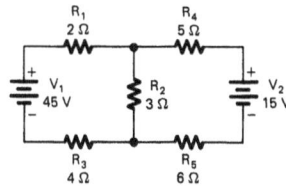

FIGURE 3-63

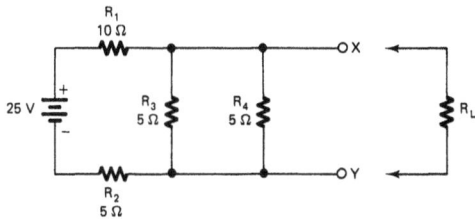

FIGURE 3-64

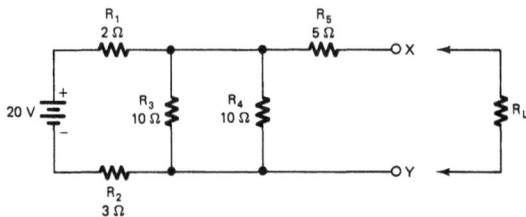

FIGURE 3-65

117. Refer to Fig. 3-61. Set up equations for loops 1, 2, and 3 of the circuit.

Solve each of the following problems by applying the super-position method.

118. Find the current flow through R_1, R_2, and R_3 in the circuit in Fig. 3-62.

119. Find the current flow through R_1, R_2, R_3, R_4, and R_5 in the circuit in Fig. 3-63.

Solve each of the following problems by applying the Thevinin equivalent-circuit method.

120. Find the Thevinin voltage (V_{TH}) and Thevinin resistance (R_{TH}) for the circuit in Fig. 3-64. Sketch the Thevinin equivalent circuit.

121. Refer to the values obtained for Fig. 3-64. Calculate the values of load current (I_L) and voltage output (V_{out}) for load resistance values of: (a) 2 Ω, (b) 3 Ω, (c) 4 Ω, and (d) 5 Ω.

122. Find the Thevinin voltage (V_{TH}) and Thevinin resistance (R_{TH}) for the circuit in Fig. 3-65. Sketch the Thevinin equivalent circuit.

123. Refer to the values obtained for Fig. 3-65. Calculate the values of load current (I_L) and output voltage (V_{out}) for load resistance values of: (a) 20 Ω, (b) 30 Ω, and (c) 50 Ω.

124. Find the values of V_{TH} and R_{TH} for the two-source circuit in Fig. 3-66.

125. Calculate I_L and V_{out} for R_L values of: (a) 10 Ω, (b) 20 Ω, (c) 30 Ω (for Fig. 3-66).

Solve the following problem by applying Norton's equivalent circuit method.

126. Find the Norton current (I_N) and Norton resistance (R_N) for the circuit in Fig. 3-67. Sketch the Norton equivalent circuit.

Solve the following problems by applying bridge-circuit simplification.

127. Find the values of V_{TH} and R_{TH} for the bridge circuit of Fig. 3-68.

128. Refer to the values obtained for Fig. 3-68. Calculate the values of load current (I_L) and output voltage for R_L values of: (a) 3 Ω, (b) 5 Ω, and (c) 8 Ω.

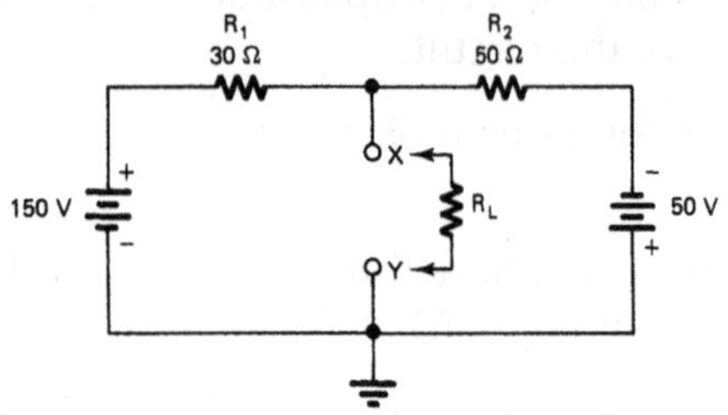

FIGURE 3-66

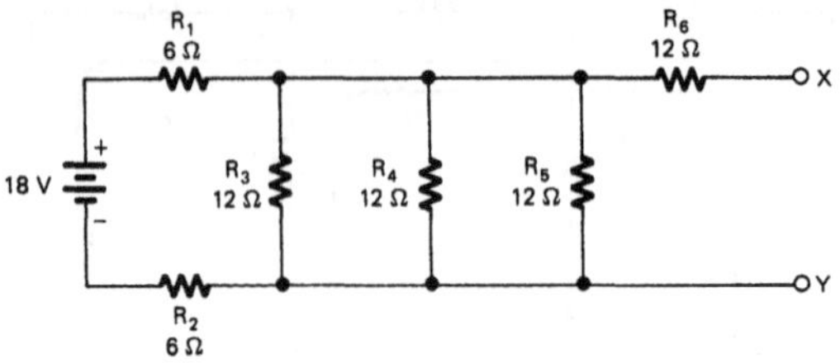

R₁ 30 Ω, R₂ 50 Ω, 150 V, R_L, X, Y, 50 V

FIGURE 3-67

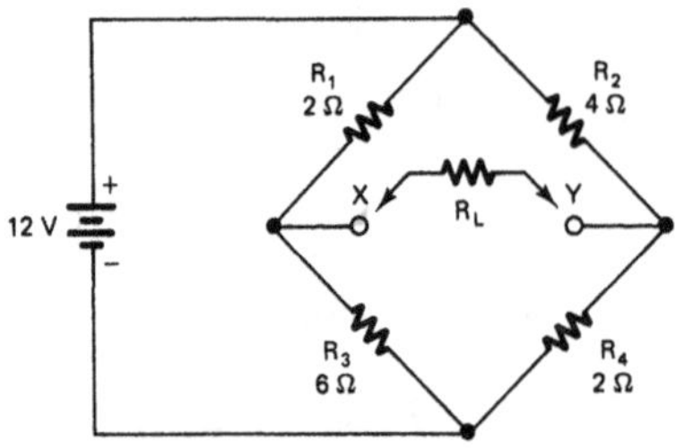

R₁ 6 Ω, R₆ 12 Ω, 18 V, R₃ 12 Ω, R₄ 12 Ω, R₅ 12 Ω, X, R₂ 6 Ω, Y

FIGURE 3-68

R₁ 2 Ω, R₂ 4 Ω, 12 V, X, R_L, Y, R₃ 6 Ω, R₄ 2 Ω

Answers

108.

R_L	I_L	V_{out}	P_{out}
0	15 A	0 V	0 W
1 Ω	12 A	12 V	144.0 W
2 Ω	10 A	20 V	200.0 W
3 Ω	8.5 A	25.7 V	218.5 W
4 Ω	7.5 A	30 V	225.0 W
5 Ω	6.67 A	33.3 V	222.1 W
6 Ω	6 A	36 V	216.0 W
7 Ω	5.45 A	38.2 V	208.4 W
8 Ω	5 A	40 V	200.0 W

109.

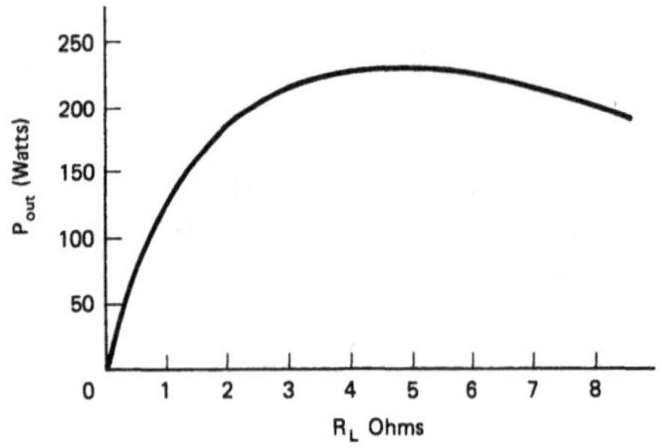

Graph: P_{out} (Watts) vs R_L Ohms; y-axis 0 to 250, x-axis 0 to 8.

110.

R_L	I_L	V_{out}	P_{out}
0 Ω	6.67 A	0 V	0 W
1 Ω	5.0 A	5.0 V	25.0 W
2 Ω	4.0 A	8.0 V	32.0 W
3 Ω	3.3 A	9.9 V	32.7 W
4 Ω	2.86 A	11.43 V	32.7 W
5 Ω	2.5 A	12.5 V	31.3 W
6 Ω	2.22 A	13.3 V	29.53 W
7 Ω	2.0 A	14.0 V	28.0 W
8 Ω	1.82 A	14.5 V	26.5 W

111. $I_1 = 6$ A   **112** $I_2 = 8$ A

113. $I_3 = 7$ A   **114** $I_4 = 2$ A

115. $I_1 = 1$ A; $I_2 = 0.2$ A;
$I_1 + I_2 = 1.2$ A

117. Loop 1 = 75 V $- I_1 R_2 -$
$V_2 - (I_1 + I_2 + I_3) R_1 = 0$
Loop 2 = 75 V $- (I_2 + I_3)$
$R_4 - I_2 R_5 - V_3 - (I_2 + I_3) R_3$
Loop 3 = 75 V $- (I_2 +$
$I_3) R_4 - I_3 R_7 - V_4 - I_3$
$R_6 - (I_2 + I_3) R_3 - (I_1$
$+ I^2 + I_3) R_1 = 0$

119. $I_{R1} = 4.99$ A;
$I_{R2} = 5.02$ A;
$I_{R3} = 4.99$ A;
$I_{R4} = 0.01$ A; $I_{R5} = 0.01$ A.

121.

R_L	I_L	V_{out}
2 Ω	0.862 A	1.72 V
3 Ω	0.695 A	2.08 V
4 Ω	0.581 A	2.32 V
5 Ω	0.5 A	2.5 V

123.

R_L	I_L	V_{out}
20 Ω	0.364 A	7.27 V
30 Ω	0.267 A	8.0 V
50 Ω	0.174 A	8.7 V

125.

R_L	I_L	V_{out}
10 Ω	0.86 A	2.57 V
20 Ω	0.64 A	3.19 V
30 Ω	0.46 A	3.69 V

127. $V_{TH} = 5$ V;
$R_{TH} = 2.83$ Ω

116. $I_1 = 1.057$ A;
$I_2 = 0.193$ A;
$I_1 + I_{2 = 1.28}$ A

118. $I_{R1} = 4.54$ A;
$I_{R2} = 2.73$ A;
$I_{R3} = 1.82$ A.

120. $V_{TH} = 3.57$ V;
$R_{TH} = 2.14$ Ω

122. $V_{TH} = 10$ V;
$R_{TH} = 7.5$ Ω

124. $V_{TH} = 75$ V;
$R_{TH} = 18.75$ Ω

126. $I_N = 1.2$ A;
$R_N = 15$ Ω

128.

R_L	I_L	V_{out}
3 Ω	0.86 A	2.57 V
5 Ω	0.64 A	3.19 V
8 Ω	0.46 A	3.69 V

EXPERIMENT 3-1

APPLICATION OF OHM'S LAW

When Ohm's law is applied to any given circuit, the computed electric values are identical to the measured electric values, if the rated values of resistors and other components are exact. Because a resistor is rated with a tolerance of ±5, ±10, or ±20%, sometimes the rated value is not the same as its measured value. This sometimes causes the computed values of current and voltage to be slightly different from the measured values.

OBJECTIVES

1. Apply Ohm's law to simple dc circuits.

2. Observe that when the measured value of voltage is doubled, the measured values of current likewise are doubled.

3. Observe that an increase in resistance brings about a decrease in current.

EQUIPMENT

Multimeter (VOM)

Resistors: 15 Ω, 100 Ω, 1 kΩ

Variable dc power supply or 1.5 V battery

Connecting wires

PROCEDURE

1. Construct the circuit in Fig. 3-1A. Use the multimeter to measure the voltage to which the variable power supply is adjusted, or use a 1.5 V battery.

2. Using the Ohm's law current formula, $I = V/R$, compute the value of I in this circuit.

3. Using the value of I just computed, adjust the VOM to the proper current range and measure I in the circuit. Measured I = _____ mA, or _____ A.

4. How did the computed value compare with the measured value? How do you explain any difference?

   _____

   _____

5. If the voltage in the circuit illustrated in Fig. 3-1A were doubled, what does Ohm's law indicate will happen to the current?

   _____

   _____

6. Double the voltage in the circuit of Fig. 3-1A and compute the new current using the Ohm's law current formula, $I = V/R$. Computed I = _____ A, or _____ mA.

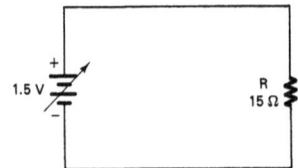

FIGURE 3-1A Simple resistive circuit.

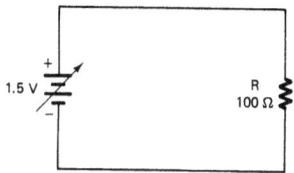

FIGURE 3-1B Resistive circuit with increased resistance.

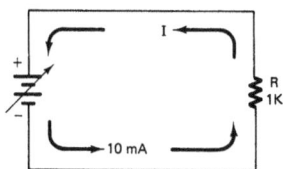

FIGURE 3-1C Circuit with 10 mA current.

7. Adjust the power supply to 3 V output. Adjust the multimeter to the proper current range and measure the new current. Measured I = _____ mA, or _____ A.

8. You can see from the foregoing procedure that as voltage increases, current increases as long as resistance remains the same. The current increases by the same factor as voltage increases. If voltage is doubled, current doubles. If voltage is increased 10 times as large, current increases 10 times.

9. Construct the circuit illustrated in Fig. 3-1B.

10. Compute the current in this circuit. Computed I = _____ A, or _____ mA.

11. Adjust the VOM to the proper current range and measure I in the circuit illustrated in Fig. 3-1B. Measured I = _____ mA, or _____ A.

12. How do computed and measured currents in steps 2 and 3 compare with the currents computed and measured in steps 10 and 11? How do you explain these differences? _____

13. Replace the 100 Ω resistor in the circuit of Fig. 3-1B with a 1 kΩ resistor. Compute the new value of current. Computed I = _____ A, or _____ mA.

14. Measure the new value of I produced when the 100 Ω resistor is replaced with the 1 kΩ resistor. Measured I = _____ A, or _____ mA.

15. How did the measured and computed currents in steps 10 and 11 compare with the currents in steps 13 and 14? How do you explain those differences?

_____

_____

16. Compute the voltage for the circuit in Fig. 3-1C that would produce 0.01 A of current. $V = I \times R$ = _____ V.

17. Adjust the power supply to the value of voltage computed in step 16 and measure the I produced by your computed value of V. Measured I = _____ mA, or _____ A.

18. If it is assumed that the value of V computed in step 16 is 10 V, Ohm's law shows that when 10 V is applied to 1 kΩ of resistance, the current is 10 mA ($I = V/R = 10/1000 = 0.01$ A, or 10 mA). What value of resistance would allow 10 times that much current if 10 V were applied? _____ Ω.

ANALYSIS

1. How is current affected by a doubled voltage when resistance remains the same?

   _____

   _____

2. How is current affected by a doubled resistance when voltage remains the same?

   _____

   _____

3. What voltage would cause 10 mA of current through a resistance of 10 Ω? _____

4. What resistance would limit the current to 1 mA when 10 V is applied? _____

5. What current would result when 1 V is applied to 1 Ω of resistance? _____

EXPERIMENT 3-2

SERIES DC CIRCUITS

A circuit is defined as the complete path or paths through which current flows. All circuits must include a voltage source as well as conductors or components through which current flows. There are three broad classifications of all electric circuits. These are *series, parallel,* and *combination* circuits.

The most easily understood circuit is the series circuit. The series circuit exhibits the following electric characteristics:

1. There is only one path for current.

2. Current has the same value everywhere in the circuit.

3. Voltage drops, when added, equal the source voltage.

4. The total resistance of the circuit is determined by means of adding the values of all resistors in the circuit.

OBJECTIVES

1. To examine the characteristics of series dc circuits.

2. To supply Ohm's law to series dc circuits.

3. To use a multimeter to make measurements in a series dc circuit.

EQUIPMENT

Multimeter (VOM)

Variable dc power supply

Resistors: 100 Ω, 220 Ω, 200 Ω

Connecting wires

PROCEDURE

1. Construct the series circuit illustrated in Fig. 3-2A.

2. In a series circuit, the total resistance is determined by means of adding the values of all of the resistors. The formula normally used in determining the total resistance of a series circuit is $R_T = R_1 + R_2 + R_3 + \ldots + R_n$. This formula shows that when the values of resistors 1 (R_1), 2 (R_2), and 3 (R_3) are added together, the result is the total resistance. Compute the total resistance of the circuit illustrated in step 1.
   $R_T =$ _____ Ω (computed)

3. Prepare the multimeter to measure resistance.

4. Disconnect the power supply. Connect the multimeter in place of the power supply. Measure and record the total resistance of the circuit:
   $R_T =$ _____ Ω (measured)

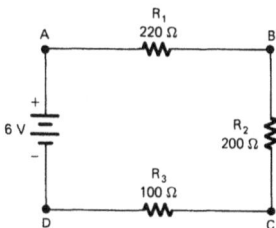

FIGURE 3-2A Series circuit.

Ohm's Law and Electric Circuits **137**

5. How does the measured value of total resistance compare with the computed value?

_____

_____

6. Disconnect the multimeter and connect the power supply to its original location in the circuit.

7. Because you now know the total resistance of the circuit and the total voltage, use Ohm's law to compute the total current. $I_T =$ _____ mA.

8. Prepare the multimeter to measure direct current. (Note: The value of current computed in step 7 will help you select the proper range).

9. Connect the multimeter at locations A, B, C, and D and record the currents at these points.

Point A: _____ mA   Point C: _____ mA

Point B: _____ mA   Point D: _____ mA

10. How do these currents compare?

_____

_____

11. Because the circuit in Fig. 3-2A is a series circuit, the current is the same through each resistor. You should have seen this in step 9. Because you know the current through each resistor and the value of each resistor, use Ohm's law to compute the voltage drop across each resistor.

V across $R_1 =$ _____ V

V across $R_2 =$ _____ V

V across $R_3 =$ _____ V

12. Prepare the multimeter to measure dc volts. Measure and record the voltage drop across each resistor.

V across $R_1 =$ _____ V

V across $R_2 =$ _____ V

V across $R_3 =$ _____ V

13. How do the voltage drops computed in step 11 compare with the voltage drops measured in step 12?

_____

_____

14. Total the voltage drops in step 12. Total voltage drop = _____ V

15. How does this total voltage drop compare with the measured source voltage?

_____

_____

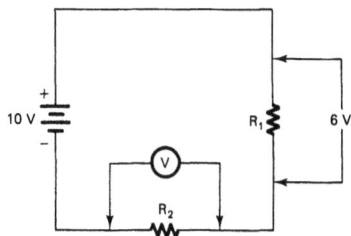

FIGURE 3-2B Computing voltage drop across resistor.

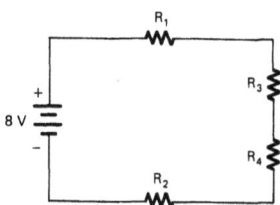

FIGURE 3-2C Four-resistor series circuit.

ANALYSIS

1. List the characteristics of a series circuit.

   _____

   _____

2. What is the voltage drop across R_2 in the circuit of Fig. 3-2B? _____

3. If three resistors valued at 6 Ω, 8 Ω, and 10 Ω, respectively, were connected in series, what would be their total resistance? $R_T =$ _____ Ω

4. In Fig. 3-2C, if 100 mA of current is flowing through R_1, how much current is flowing through R_2, R_3, and R_4?

5. How many paths for current are illustrated in the circuit of Fig. 3-2C? _____

6. What is the total voltage drop in the circuit illustrated in Fig. 3-2C? _____

7. If three resistors valued at 10 Ω each were connected in series to a voltage source, what portion of the source voltage would appear across each resistor? _____

EXPERIMENT 3-3

PARALLEL DC CIRCUITS

Parallel circuits are circuits that provide two or more paths for current. Most of the light fixtures and wall outlets in your home are connected in parallel circuits. Many of the circuits that cause your radios and television sets to operate are parallel circuits.

The characteristics of parallel circuits are as follows:

1. Two or more paths are provided for current.

2. The voltage across each path or branch is the same.

3. The sum of the currents in each path or branch equals the total current.

4. The total resistance (R_T) is computed with the following formula:

$$R_T = \cfrac{1}{\cfrac{1}{R_1} + \cfrac{1}{R_2} + \cfrac{1}{R_3} + \ldots + \cfrac{1}{R_n}}$$

where $R_1, R_2, \ldots, R_n$ are the resistances of branches $1, 2, \ldots, n$. This equation is the same as

$$\frac{1}{R_T} = \frac{1}{R_T} + \frac{1}{R_2} + \frac{1}{R_3} + \ldots + \frac{1}{R_n}$$

OBJECTIVES

1. To examine the characteristics of parallel dc circuits.

2. To apply Ohm's law to parallel dc circuits.

3. To use a multimeter to make measurements in a parallel dc circuit.

EQUIPMENT

Multimeter (VOM)

Variable dc power supply

Resistors: 10 Ω, 15 Ω, 22 Ω

Connecting wires

PROCEDURE

1. Construct the circuit illustrated in Fig. 3-3A. Use either a 1.5 V battery or a power supply.

2. How many different paths are provided for current? Is this a parallel or series circuit? _____

3. What is the voltage across each path? _____ V

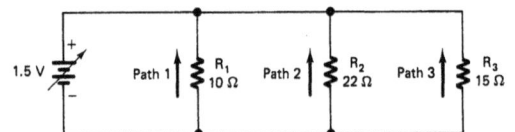

FIGURE 3-3A Three-branch parallel circuit.

4. Compute the current in each path and the total current for the circuit:

Path 1 = _____ mA

Path 2 = _____ mA

Path 3 = _____ mA

Total = _____ mA

5. Compute the total resistance for the circuit illustrated in Fig. 3-3A. $R_T =$ _____ Ω.

6. Prepare the multimeter to measure dc. Measure and record the currents through path 1 (R_1), path 2 (R_2), and path 3 (R_3), and the total current in the circuit. (The meter should be connected as illustrated in Fig. 3-3B for each of these measurements. Thus the current in the path will correctly flow through the meter.)

$I_T =$ _____ mA

$I_1 =$ _____ mA

$I_2 =$ _____ mA

$I_3 =$ _____ mA

7. How do the values of current in step 4 compare with the values of current in step 6?

_____

_____

8. Disconnect the multimeter from the circuit and restore the circuit to its original condition as shown in Fig. 3-3A.

9. Prepare the multimeter to measure dc volts. Measure and record the voltage across each path of the circuit illustrated in Fig. 3-3A.

Path 1 = _____ V

Path 2 = _____ V

Path 3 = _____ V

10. How do these voltages compare with each other?

_____

_____

11. Disconnect the power supply. Prepare the multimeter to measure resistance. Connect the multimeter in place of the power supply and measure the total resistance of the circuit: $R_T =$ _____ Ω.

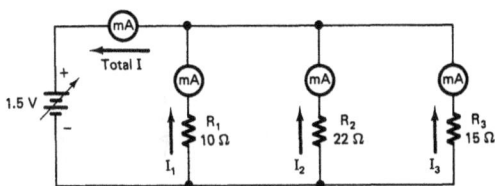

FIGURE 3-3B Connection of VOM to measure direct current.

12. How does the measured total resistance in step 11 compare with the computed total resistance in step 5?

_____

_____

ANALYSIS

1. List the characteristics of a parallel circuit.

_____

_____

2. What is the current through R_3 in the circuit illustrated in Fig. 3-3C? _____

3. What is the total current in the circuit of Fig. 3-3D?

_____

4. What is the voltage across R_2 in the circuit of Fig. 3-3D? _____

5. What is the total resistance of the circuit shown in Fig. 3-3D? _____

6. If the resistance of paths of a parallel circuit is equal, how will the current in one path compare with the current in other paths?

_____

_____

7. In the circuit of Fig. 3-3E, indicate where the multimeter should be placed to measure total current. Use an X to indicate placement.

8. In the circuit of Fig. 3-3E, indicate where the multimeter should be placed to measure current through R_2. Use a large dot to show the placement of the meter.

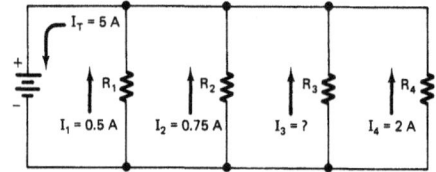

FIGURE 3-3C Circuit with unknown branch current.

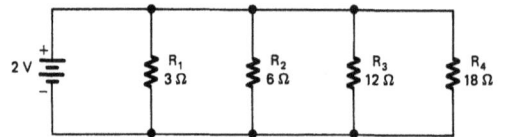

FIGURE 3-3D Circuit with unknown total current.

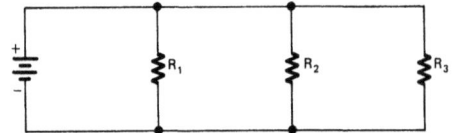

FIGURE 3-3E Current measurement.

EXPERIMENT 3-4

COMBINATION DC CIRCUITS

Combination circuits are the most commonly used circuits in electronics. The name is derived from the fact that the circuit is a combination of series and parallel circuits and that it exhibits the characteristics of both. When working with a combination circuit, you must treat its series portion as a series circuit and its parallel portion as a parallel circuit. You must also understand the influence of one portion of a combination circuit on the other portions.

All current, voltage, and power computations and measurements for a combination circuit are made in the same way as in previous circuits. The total resistance of a combination circuit is determined by means of first determining the series resistance, then the parallel resistance, and finally the sum.

OBJECTIVES

1. To examine the characteristics of a combination dc circuit.

2. To apply Ohm's law to combination dc circuits.

3. To use a multimeter to make measurements in a combination dc circuit.

EQUIPMENT

Multimeter (VOM)

Variable dc power supply or 6 V battery

Resistors: 10 Ω, 15 Ω, 22 Ω

Connecting wires

PROCEDURE

1. Construct the circuit shown in Fig. 3-4A.

2. Prepare the multimeter to measure dc voltage. Measure and record the voltages across R_1, R_2, and R_3.

   $V_{R1} =$ _____ V

   $V_{R2} =$ _____ V

   $V_{R3} =$ _____ V

3. How did the voltage across R_2 compare with the voltage across R_3?

   _____

   _____

4. How does the sum of the voltage across R_1 and the voltage across R_2 or R_3 compare with the source voltage?

   _____

   _____

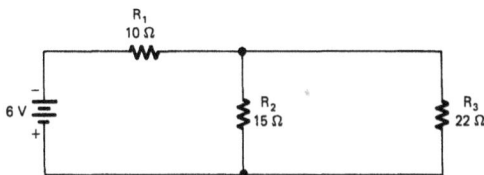

FIGURE 3-4A Series-parallel combination circuit.

5. Prepare the multimeter to measure direct current. Measure and record the currents through R_1, R_2, and R_3 and the total current.

    $I_T = $ _____

    $I_{R1} = $ _____

    $I_{R2} = $ _____

    $I_{R3} = $ _____

6. How does the total current compare with the current through R_1?

    _____

    _____

7. How does the sum of the currents through R_2 and R_3 compare with the total current?

    _____

    _____

8. How does the sum compare with the current through R_1?

    _____

    _____

9. Compute the total resistance for the circuit shown in Fig. 3-4A. (Remember to add the series resistance to the parallel resistance.)
    $R_T = $ _____ Ω.

10. Disconnect the power supply from the circuit in Fig. 3-4A. Prepare the multimeter to measure resistance. Connect the multimeter where the power supply was connected. Measure and record the total resistance of the circuit.
    $R_T = $ _____ Ω.

11. How did the measured resistance in step 10 compare with the computed resistance in step 9?

    _____

    _____

12. Using Ohm's law, compute the total current for the circuit in step 1. $I_T = $ _____ A.

13. How does the current computed in step 12 compare with the current measured in step 5?

    _____

    _____

14. How does the total current in this circuit compare with the current through R_1?

    _____

    _____

15. Compute the total parallel resistance of R_2 and R_3.
    Parallel resistance of R_2 and R_3 = _____ Ω.

16. In step 7, you found that the sum of the currents through R_2 and R_3 equals the total current. Using the value of total current computed in step 12 and the parallel resistance computed in step 15, compute the voltage across the parallel resistors R_2 and R_3. V across R_2 and R_3 = _____ V.

17. How does this voltage compute with the measured values in step 2?

_____

_____

18. Using the total current computed in step 12, compute the voltage across R_1. V_{R1} = _____ V.

19. How does the computed voltage in step 18 compare with the measured voltage in step 2?

_____

_____

20. How does the sum of the computed voltages across R_1 and R_2 and R_3 compare with the source voltage?

_____

_____

ANALYSIS

1. How does the total current of a combination circuit compare with the current through its series components?

_____

_____

2. How does the sum of the currents in the parallel paths of a combination circuit compare with the total current of the circuit?

_____

_____

3. How does the sum of the currents in the parallel paths of a combination circuit compare with the current through the series components?

_____

_____

4. How does the sum of the voltages across the series components and the parallel paths of a combination circuit compare with the source voltage?

_____

_____

5. If you were to subtract the voltage across the parallel paths of a combination circuit from the source voltage, how would the results compare with the voltage across the series components of the circuit?

_____

_____

6. If you were to subtract the sum of the voltages across the series components of a combination circuit from the source voltage, how would the results compare with voltage across the parallel paths of the circuit?

_____

_____

7. If you were to add the sum of the resistance of the series components of a combination circuit to the total parallel resistance of the circuit, how would the results compare with the total resistance of the circuit?

_____

_____

EXPERIMENT 3-5

POWER IN DC CIRCUITS

Electric power is probably the electric quantity most familiar to most people. All our electric utility bills are computed relative to the amount of power consumed in a given amount of time. All electric appliances have a specified power rating, which indicates how economically they may be operated. Knowledge of electric power is important to anyone who lives in this age of energy conservation.

Electric power is consumed each time a voltage causes current flow. Power generally appears in the form of heat, light, or motion. The basic unit of measurement for electric power is the watt. The milliwatt and kilowatt also are commonly used as units of measurement of power. Power is the product of current and voltage in a circuit. That means power (in watts) equals current (in amperes) multiplied by voltage (in volts). Power can be computed when any two of the following electric quantities are known: voltage (V), current (I), or resistance (R). The three basic formulas used in computing power in watts are as follows:

$$P = IV \quad P \;=\; I^2R \quad P = \frac{V^2}{R}$$

where *P, V, I,* and *R* are in watts, volts, amperes, and ohms, respectively.

Many electric components have an electric power rating. Resistors are an excellent example, having common power ratings of ⅛ Ω, ¼ Ω, ½ Ω, 1 Ω, and so on. This rating indicates the maximum current that can pass through the resistor without damaging the component.

OBJECTIVES

1. To calculate power values in dc circuits by means of applying the basic power formulas.

2. To use a multimeter to make measurements to determine power in dc circuits.

EQUIPMENT

Multimeter (VOM)

Variable dc power supply or 6 V battery

6 V lamp and socket

Resistors: 10 Ω, 100 Ω, 220 Ω, 1 kΩ

Connecting wires

SPST switch

PROCEDURE

1. Construct the circuit shown in Fig. 3-5A.

2. Adjust the multimeter to measure current on the highest range.

3. Connect the multimeter to the circuit illustrated in Fig. 3-5A. Close the SPST switch and measure and record the current. _____ mA, or _____ A.

4. Open the SPST switch, disconnect the multimeter, and restore the circuit to its original state, illustrated in Fig. 3-5A.

5. Prepare the multimeter to measure dc voltage.

6. Connect the multimeter to measure the voltage drop across the lamp illustrated in the circuit. Close the switch and record the voltage across the lamp. _____ V.

7. You now have measured current through the lamp and the voltage across the lamp. Compute the power converted by the lamp. _____ W.

8. The power converted increases when voltage increases and resistance remains the same or when resistance decreases and voltage remains the same. Observe the relation between current, voltage, resistance, and power as you solve the following problems.

   a. $R = 10\ \Omega$, $I = 2$ A, $P =$ _____ W,
      $V =$ _____ V

   b. $V = 100$ V, $I = 100$ mA, $P =$ _____ W,
      $R =$ _____ Ω

   c. $I = 4$ A, $R = 5\ \Omega$, $P =$ _____ W,
      $V =$ _____ V.

   d. $R = 20\ \Omega$, $V = 10$ V, $P =$ _____ W,
      $I =$ _____ A

   e. $I = 1.5$ A, $V = 90$ V, $P =$ _____ W,
      $R =$ _____ Ω

9. Construct the circuit shown in Fig. 3-5B.

10. Set the multimeter to measure direct current in the 12 mA range or equivalent.

11. Connect the multimeter to the circuit to measure current, close the SPST switch, and record the current indicated by the meter. _____ mA.

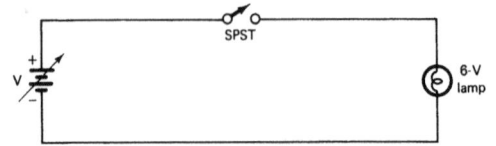

FIGURE 3-5A Simple 6 V lamp circuit.

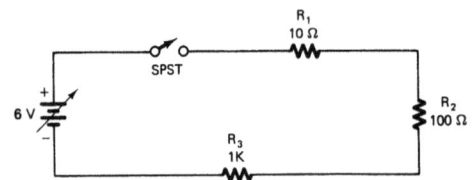

FIGURE 3-5B Circuit with three resistors.

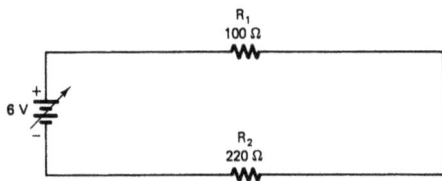

FIGURE 3-5C Two-resistor circuit.

12. Compute the power converted by the circuit illustrated in Fig. 3-5B. _____ W.

13. Open the switch, disconnect the multimeter, and alter the circuit to the one in Fig. 3-5C.

14. Set the multimeter to the 12 V dc range. Measure and record the voltage across R_1, R_2, and R_3. Be sure to observe proper meter polarity.

   V across R_1 = _____ V

   V across R_2 = _____ V

   V across R_3 = _____ V

15. Compute the power converted by R_1 and R_2.

   Computed power for R_1 = _____ Ω

   Computed power for R_2 = _____ Ω

16. How does the power converted by R_1 compare with the power converted by R_2? Why do they differ?

   _____

   _____

ANALYSIS

1. How much current flows through a 100 W light bulb when 100 V is connected to it? _____

2. How much power is converted by a 12 V lamp with a filament resistance of 2 Ω? _____

3. What is the maximum current that could safely pass through a 1000 Ω, 1 W resistor? _____

4. What power is converted by a 5 kΩ resistor that has 1 mA of current through it? _____

5. What power is converted by a light bulb that drops 6 V because of its 10 Ω filament?

   _____

6. What is the maximum voltage drop that could be developed by a 1.2 kΩ, ½ W resistor? _____

EXPERIMENT 3-6

VOLTAGE-DIVIDER CIRCUITS

A very important type of electronic circuit is the voltage-divider circuit. This type of circuit in its simplest form consists of two or more resistors connected in series with a power source. A voltage-divider circuit allows one to obtain several different voltages from one power source simply by means of changing values of resistance. Such circuits are used extensively in electronic equipment. The power-supply system of a transistor radio is a good example of a system that uses voltage-divider circuits to obtain various voltages to operate the radio circuitry.

OBJECTIVES

1. To study the potentiometer as a voltage divider.

2. To calculate resistance values to design a voltage-divider circuit.

3. To use a multimeter to make measurements in a voltage-divider circuit.

EQUIPMENT

Variable dc power supply

Multimeter (VOM)

Potentiometer: 10 KΩ

Resistors: 1.2 kΩ, 470 Ω

Connecting wires

PROCEDURE

Section A: Potentiometer Voltage Division

1. Obtain a 200 Ω potentiometer (see Fig. 3-6A).

2. Measure the resistance across the two outer terminals of the potentiometer. R_{AC} = _____ Ω. Does rotating the shaft have any effect on this resistance reading? _____

3. Record the following ohmmeter readings:
   a. R_{AB} with shaft turned fully clockwise:
      R_{AB} = _____ Ω.
   b. R_{AB} with shaft turned fully counterclockwise:
      R_{AB} = _____ Ω.
   c. R_{BC} with shaft turned fully counterclockwise:
      R_{BC} = _____ Ω.
   d. R_{BC} with shaft turned fully clockwise:
      R_{BC} = _____ Ω.

4. Construct the circuit shown in Fig. 3-6B by applying 6 V to the outer terminals of the potentiometer (point A to point C).

5. Adjust V_{BC} to 1 V by means of rotating the potentiometer shaft. Turn off the power supply and

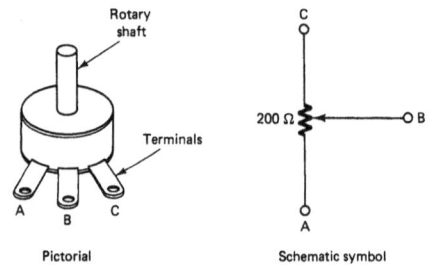

FIGURE 3-6A Potentiometer.

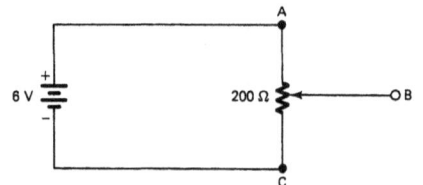

FIGURE 3-6B Potentiometer circuit.

V_{BC}	R_{BC}
1 V	
2 V	
4 V	
6 V	

FIGURE 3-6C Potentiometer voltage division.

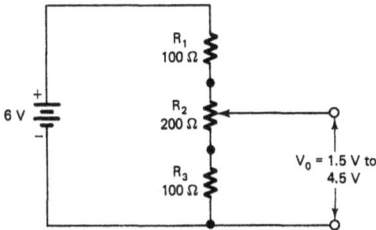

FIGURE 3-6D Voltage-divider circuit.

remove the wire at point *C*. Measure resistance R_{BC}. Reconnect the wire. Record the value of R_{BC} in Fig. 3-6C.

6. Adjust V_{BC} to 2 V. Turn off the power supply and measure V_{BC} as in step 5. Record R_{BC} in Fig. 3-6C.

7. Increase V_{BC} to 4 V and then 6 V and complete Fig. 3-6C.

Section B: Voltage-Divider Design

1. Refer to the circuit shown in Fig. 3-6D.

2. Determine the values of R_1 and R_3 needed to produce the output voltage (V_{out}) across the movable potentiometer contact and (–) terminal of the power supply to vary from 2V to 4.5V. This represents a variable dc voltage output determined by the values of R_1 and R_3.

3. Use the voltage across R_2 and its resistance to determine the current that will flow in the circuit, as follows:

$$I = \frac{V_{R2}}{R_2} = \underline{\hspace{1cm}} A$$

4. Determine the values of R_1 and R_3 on the basis of the current calculated in step 3 and the voltage drops across R_1 and R_3. $R_1 = \underline{\hspace{1cm}} \Omega$; $R_3 = \underline{\hspace{1cm}} \Omega$.

5. Construct the circuit shown in Fig. 3-6D using your calculated values of R_1 and R_3.

6. Measure the voltage range of V_{out} as the potentiometer shaft is rotated. $V_{out} = \underline{\hspace{1cm}}$ V to $\underline{\hspace{1cm}}$ V.

ANALYSIS

1. From the data of Fig. 3-6C, verify that the voltage drop across a resistance varies directly with the amount of resistance.

$$\frac{2V}{4V} = \frac{R_{BC} \text{ with 2 Voutput}}{R_{BC} \text{ with 4 Voutput}} = \underline{\hspace{1cm}} \frac{\Omega}{\Omega}$$

2. The voltage-division equation is stated as follows: $V_{out} = (R_{out}/R_T) \times V_T$, where V_{out} is the output voltage, R_T is the total resistance in the branch circuit,

R_{out} is the output resistance, and V_S is the source voltage (Fig. 3-6E). Verify this equation using the data of Fig. 3-6C at $V_{BC} = 4$ V. _____

3. Two transistors are to be supplied the following dc voltages and currents: 5 V at 10 mA and 3 V at 5 mA. Use the circuit in Fig. 3-6F to determine the values of R_1, R_2, and R_3 required for this voltage-divider circuit.

   $R_1 =$ _____ Ω

   $R_2 =$ _____ Ω

   $R_3 =$ _____ Ω

4. Determine the minimum power rating of each resistor in the voltage divider of Fig. 3-6F.
   $P_{R1} =$ _____ W; $P_{R2} =$ _____ W;
   $P_{R3} =$ _____ W.

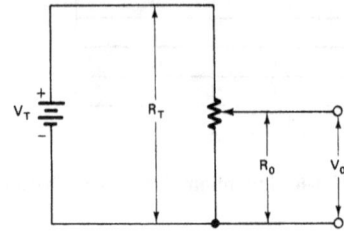

FIGURE 3-6E Circuit for voltage division.

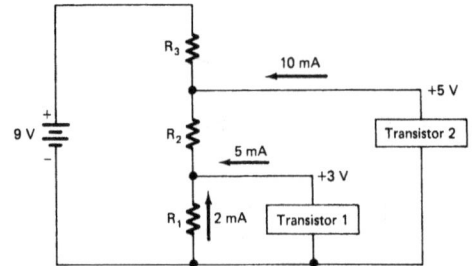

FIGURE 3-6F Voltage-divider circuit as transistor supply.

KIRCHHOFF'S VOLTAGE LAW

Kirchhoff's voltage law is basic to the understanding of electric circuitry. We may express this voltage law in the two following ways:

1. The sum of the voltage drops in any single current path of a circuit is equal to the voltage applied to that path.

2. The algebraic sum of the voltage drops and voltage sources of a current path of a circuit is equal to zero.

OBJECTIVES

1. To study the application of Kirchhoff's voltage law for circuits with one voltage source and those with more than one voltage source.

2. To use a multimeter to measure voltages for applying Kirchhoff's voltage law.

EQUIPMENT

Variable dc power supply

Multimeter (VOM)

Resistors: 470 Ω, 1 kΩ, 1.2 kΩ, 2 kΩ, 2.2 kΩ, 4.7 kΩ

Connecting wires

Dry cells (2): 1.5 V

PROCEDURE

Section A: Single Voltage Source

1. Construct the circuit in Fig. 3-7A. Apply 10 V.

2. Starting at point A, move in a counterclockwise (ccw) direction around the closed loop and measure each voltage drop indicated. (The first subscript letter designates the placement of the negative meter lead). V_{AB} = _____ V; V_{BC} = _____ V; V_{CD} = _____ V; V_{DE} = _____ V.

3. Verify Kirchhoff's voltage law:

$$V_{AB} + V_{BC} + V_{CD} + V_{DE} = V_S$$

_____ V + _____ V + _____ V + _____ V = 10 V

4. Set up the Kirchhoff's law equation for loop 1 shown in Fig. 3-7A. _____

Section B: Multiple Voltage Sources

1. Construct the circuit shown in Fig. 3-7B. Use 10 V from the power supply and two dry cells as power sources. Check each dry cell under load to assure that they are good.

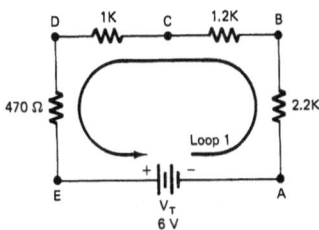

FIGURE 3-7A Circuit with single voltage source.

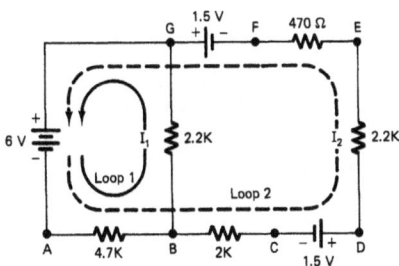

FIGURE 3-7B Circuit with multiple voltage sources.

2. Measure each voltage present in the circuit. (The first subscript letter designates the placement of the negative meter lead).

$V_{AB} = $ _____ V; $V_{BC} = $ _____ V

$V_{CD} = $ _____ V; $V_{DE} = $ _____ V

$V_{EF} = $ _____ V; $V_{FG} = $ _____ V

$V_{AG} = $ _____ V; $V_{BG} = $ _____ V

3. Set up the equation for each current loop.

Equation for loop 1: _____

Equation for loop 2: _____

4. Combine the equations and solve for each current value.

$I_1 = $ _____ mA

$I_2 = $ _____ mA

$I_1 + I_2 = $ _____ mA

ANALYSIS

1. State Kirchhoff's voltage law.

_____

_____

2. Why is it important to be able to apply Kirchhoff's voltage law?

_____

_____

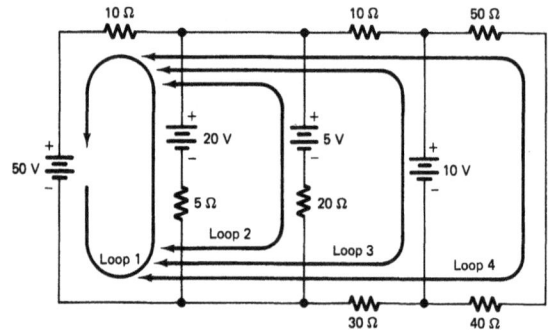

FIGURE 3-7C Four-loop circuit.

3. Set up the equation for each current loop of the circuit in Fig. 3-7C.

Equation for loop 1: _____

Equation for loop 2: _____

Equation for loop 3: _____

Equation for loop 4: _____

4. Set up the equations for each current loop of the circuit in Fig. 3-7D.

Equation for loop 1: _____

Equation for loop 2: _____

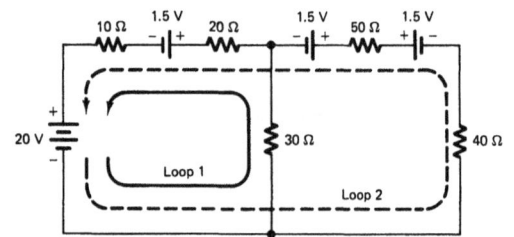

FIGURE 3-7D Two-loop circuit.

KIRCHHOFF'S CURRENT LAW

Another basic electric law involves the distribution of current in a circuit. We know that the amount of current that flows in a circuit is determined by the amount of voltage applied and the total resistance of the circuit. Kirchhoff's current law is expressed as follows: *The sum of the currents flowing to a point in a circuit is equal to the sum of the current flowing away from that point.* This very simple statement is fundamental to the understanding of electronic circuits.

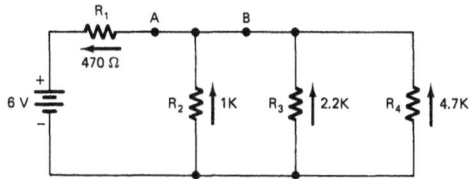

FIGURE 3-8A Circuit application of Kirchhoff's current law.

OBJECTIVES

1. To study the application of Kirchhoff's current law for a combination circuit.

2. To use a multimeter to measure currents to apply Kirchhoff's current law.

EQUIPMENT

Variable dc power supply

Multimeter (VOM)

Resistors: 470 Ω, 1 kΩ, 2.2 kΩ, 4.7 kΩ

Connecting wires

PROCEDURE

1. Calculate the values of total current (I_{R1}) and the currents through each of the resistors (I_{R2}, I_{R3}, I_{R4}) for the circuit shown in Fig. 3-8A:
   I_{R1} = —————— mA; I_{R2} = —————— mA; I_{R3} = —————— mA; I_{R4} = —————— mA.

2. Construct the circuit of Fig. 3-8A. Adjust the power supply to 10 V. Measure each of the following currents with a meter: I_{R1} = —————— mA; I_{R2} = —————— mA; I_{R3} = —————— mA; I_{R4} = —————— mA.

3. Verify Kirchhoff's current law by looking at the value of current flowing into point *A* of the circuit. (Use measured values.)

   $I_{R1} = I_{R2} + I_{R3} + I_{R4}$ = —————— + —————— + —————— = —————— mA.

4. Measure the current through point *B* of the circuit. Current = —————— mA.

5. Verify Kirchhoff's current law as follows using your measured values:

   $I_B = I_{R3} + I_{R4}$ = —————— + —————— = —————— mA.

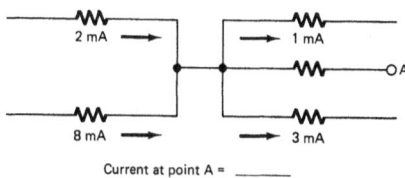

Current at point A = ————

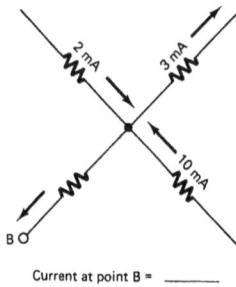

Current at point B = ————

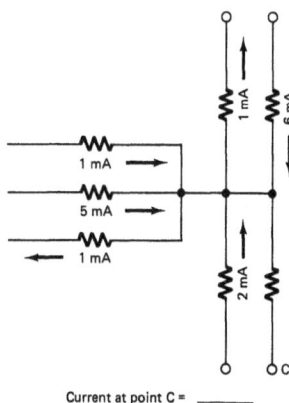

Current at point C = ————

FIGURE 3-8B

ANALYSIS

1. State Kirchhoff's current law.

   _____

   _____

2. What are some applications of Kirchhoff's current law?

   _____

   _____

3. Solve the problems of Fig. 3-8B by applying Kirchhoff's current law.

4. Calculate the percentage difference between your measured and calculated values of total current (steps 1 and 2).

% Difference = [(calculated value – measured value) ÷ calculated value]

SUPERPOSITION METHOD

Experiment 3-7 deals with the use of Kirchhoff's voltage law to solve electric circuit problems. Another method that may be used to solve problems with circuits that have more than one voltage source is called the *superposition method*. A circuit must be reduced to as many individual circuits as there are voltage sources. If a circuit has two voltage sources, two separate circuits derived from the original circuit must be manipulated. In this method the currents produced by each voltage source are calculated individually. The direction of each current must be labeled. When the currents from each voltage source have been determined, they may be combined to find the current through each component of the circuit.

OBJECTIVES

1. To study the superposition method for determining the currents in a multiple-source circuit.

2. To use a multimeter to measure currents for applying the superposition method.

3. To compare superposition with Kirchhoff's voltage law method.

EQUIPMENT

Variable dc power supply

Multimeter (VOM)

Dry cell: 1.5 V

Resistors: 1 kΩ, 2 kΩ, 2.2 kΩ

Connecting wires

1. Using the superposition method, calculate the currents through each resistor of the circuit in Fig. 3-9A. Label these values on the circuits sketched in the figure. Draw arrows to indicate the directions of current.

2. Construct the original circuit shown in Fig. 3-9A. Apply 10 V from the variable dc power supply. Use a dry cell for the other power source. Check the dry cell to determine its condition.

3. Measure the current through each resistor:
$I_{R1} =$ —————— mA; $I_{R2} =$ —————— mA; $I_{R3} =$ —————— mA.

4. Apply Kirchhoff's voltage law to calculate each of the currents in the circuit shown in Fig. 3-9A.

Equation for loop 1: ——————

Equation for loop 2: ——————

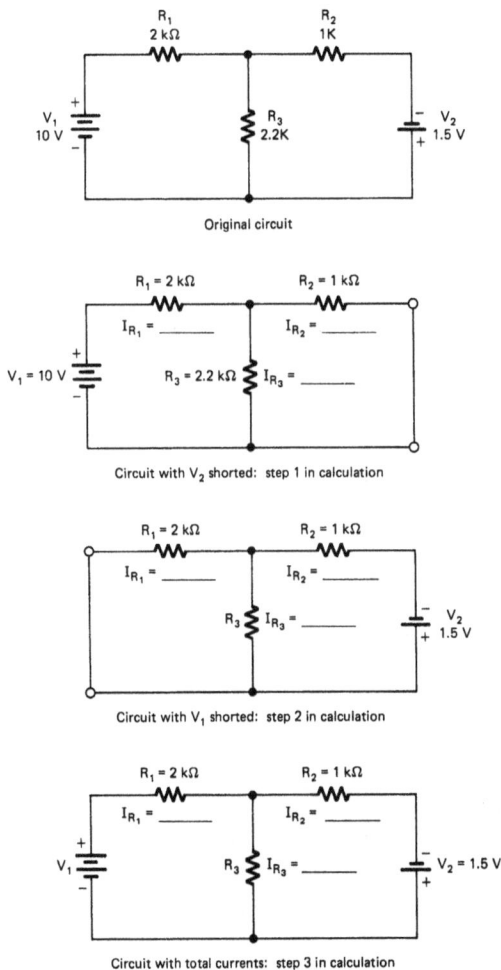

Original circuit

Circuit with V_2 shorted: step 1 in calculation

Circuit with V_1 shorted: step 2 in calculation

Circuit with total currents: step 3 in calculation

FIGURE 3-9A Illustration of superposition method.

Combine the equations and solve for I_1, I_2, and $I_1 + I_2$.

$I_1 = $ —————— mA

$I_2 = $ —————— mA

$I_1 + I_2 = $ —————— mA

ANALYSIS

1. Which method, Kirchhoff's law or superposition, do you prefer for solving multiple-source problems? Why?

    _____

    _____

2. Why is it necessary to use Kirchhoff's law or the superposition method to solve multiple-source problems?

    _____

    _____

3. Determine the percentage difference between your measured and calculated values of I_{R1} (steps 1 and 3). % Difference = —————— %.

4. State the superposition method.

    _____

    _____

THEVININ EQUIVALENT CIRCUITS

Many electronic circuit applications necessitate use of equivalent circuits. Equivalent circuit methods are used to reduce complex circuits to much simpler forms for circuit design or servicing work. One such equivalent circuit method is Thevinin's theorem. *Thevinin's theorem is expressed as follows: A complex circuit may be reduced to an equivalent voltage source and an equivalent series resistance.* The equivalent voltage is called the Thevinin voltage (V_{TH}), and the equivalent series resistance is called the Thevinin resistance (R_{TH}). The equivalent circuit has the same electric characteristics as the circuit from which it was derived.

OBJECTIVES

1. To study the procedure used to make an equivalent circuit using Thevinin's theorem for a single voltage source problem and to solve a problem with two voltage sources.

2. To use a multimeter to make measurements for applying Thevinin's theorem.

EQUIPMENT

Variable dc power supply or 6 V battery

Multimeter (VOM)

Dry cell: 1.5 V

Resistors: 240 Ω, 470 Ω, 1 kΩ, 1.2 kΩ, 2.2 kΩ

Connecting wires

PROCEDURE

Section A: Single Voltage Source

1. Calculate the Thevinin equivalent voltage V_{TH} and Thevinin equivalent resistance R_{TH} for the circuit shown in Fig. 3-10A. Label these values on the diagram.

2. Construct the circuit shown in Fig. 3-10A and apply 6 V from your power supply or a 6 V battery.

3. Measure the voltage across the output terminals (points X and Y). This is the Thevinin equivalent voltage. $V_{TH} =$ _____ V.

4. Place a 2.2 kΩ resistor across the output terminals (points X and Y).

5. Measure the output voltage (V_{out}) across points X and Y. $V_{out} =$ _____ V.

6. Compute the output voltage using the values of V_{TH} and R_{TH} from Fig. 3-10A. $V_{out} =$ _____ V.

7. Remove the 2.2 kΩ resistor and disconnect the circuit from the power source.

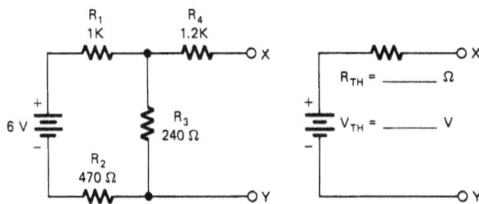

FIGURE 3-10A Circuit with single voltage source.

8. Place a wire across the points where the power supply was connected to the circuit. Use an ohmmeter to measure the Thevinin equivalent resistance (R_{TH}). R_{TH} = ————— Ω.

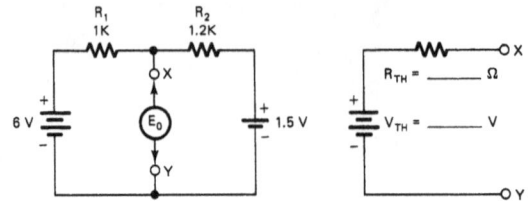

FIGURE 3-10B Circuit with two voltage sources.

Section B: Multiple Voltage Source

1. Calculate V_{TH} for the circuit shown in Fig. 3-10B. Label these values on the diagram.

2. Construct the circuit of Fig. 3-10B using 6 V from the power supply or a battery and a 1.5 V dry cell as the other power source.

3. Measure the voltage across the output terminals (points X and Y). V_{TH} = ————— V.

4. Place a 470 Ω resistor across the output terminals and measure the output voltage.
V_{out} = ————— V.

5. Compute the output voltage using the values of V_{TH} and R_{TH} from step 1. V_{out} = ————— V.

6. Remove the 470 Ω load resistor and disconnect the circuits from the power sources.

7. Place a wire across the points where the power sources were connected to the circuit. Use an ohmmeter to measure the Thevinin resistance.
R_{TH} = ————— Ω.

ANALYSIS

1. State Thevinin's theorem.

————————————————————————

————————————————————————

2. Compare the percentage difference between measured and calculated values for the following:
   a. V_{TH} (steps 1 and 3, section A): ————— %
   b. R_{TH} (steps 1 and 8, section A): ————— %
   c. V_{out} (steps 4 and 5, section B): ————— %

3. Why is it advantageous to use a Thevinin equivalent circuit?

————————————————————————

————————————————————————

4. Using the equivalent circuit values for R_{TH} and V_{TH}, calculate the output voltage for the following loads connected across X and Y:
   a. R_L = 150 Ω, V_{out} = ————— V
   b. R_L = 2 kΩ, V_{out} = ————— V
   c. R_L = 12 kΩ, V_{out} = ————— V

5. Using the equivalent circuit values for R_{TH} and V_{TH} of section B, compute the output voltage for the following loads:

a. $R_L = 10\ \Omega$, $V_{out} =$ —————— V

b. $R_L = 18\ k\Omega$, $V_{out} =$ —————— V

c. $R_L = 200\ k\Omega$, $V_{out} =$ —————— V

EXPERIMENT 3-11

NORTON EQUIVALENT CIRCUITS

Another method to reduce a complex circuit into a simple equivalent circuit is application of Norton's theorem. This method is used to reduce a circuit to an equivalent current source and a parallel resistance. The source current is called the *Norton current* (I_N), and the parallel resistance is called the *Norton resistance* (R_N). This type of circuit is similar to the Thevinin equivalent circuit except it deals with a current source and parallel resistance rather than a voltage source and a series resistance.

OBJECTIVES

1. To study the procedure used to make an equivalent circuit using Norton's theorem and to observe the similarity between this method and the Thevinin equivalent circuit method.

2. To use a multimeter to make measurements for applying Norton's theorem.

EQUIPMENT

Variable dc power supply or 6 V battery

Multimeter (VOM)

Resistors: 100 Ω, 220 Ω, 240 Ω, 470 Ω, 1 kΩ, 1.2 kΩ, 2.2 kΩ, 4.7 kΩ

Connecting wires

PROCEDURE

1. Calculate the values of Norton equivalent resistance (R_N) and Norton equivalent current (I_N) for the circuit in Fig. 3-11A. Label the values of I_N and R_N and the direction of I_N on the diagram.

2. Using the formula $I_L = (I_N R_N) \div (R_N + R_L)$, determine the values of load current for each of the following values of load resistance:
   a. $R_L = 240\ \Omega$, $I_L =$ ———— mA
   b. $R_L = 4.7\ k\Omega$, $I_L =$ ———— mA
   c. $R_L = 100\ \Omega$, $I_L =$ ———— mA
   d. $R_L = 220\ \Omega$, $I_L =$ ———— mA

3. Construct the circuit shown in Fig. 3-11A. Place a shorting wire across terminals X and Y, then apply 6 V.

4. Measure the current at point A. This is the Norton equivalent current (I_N). $I_N =$ ———— mA.

5. Remove the shorting wire and measure the current (I_L) that flows through each of the four load resistors of step 2 as they are placed across the X and Y terminals:
   a. $R_L = 240\ \Omega$, $I_L =$ ———— mA
   b. $R_L = 4.7\ k\Omega$, $I_L =$ ———— mA

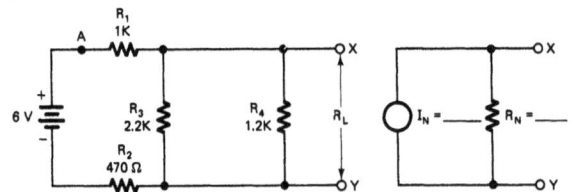

FIGURE 3-11A Circuit for application of Norton's theorem.

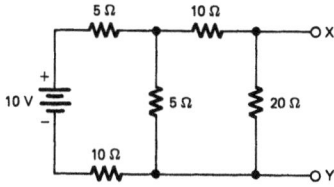

FIGURE 3-11B Combination circuit problem.

c. $R_L = 100 \ \Omega$, $I_L = $ ———— mA

d. $R_L = 220 \ \Omega$, $I_L = $ ———— mA

ANALYSIS

1. Determine the percentage difference of the measured and calculated values of Norton equivalent current (I_N) from steps 1 and 4. % Difference = ———— %.

2. State Norton's theorem.

_____

_____

3. How do Norton's theorem and Thevinin's theorem relate to one another?

_____

_____

4. Determine the values of I_N and R_N for the circuit in Fig. 3-11B. $R_N = $ ———— Ω; $I_N = $ ———— A.

EXPERIMENT 3-12

MAXIMUM POWER TRANSFER

In this experiment you will study maximum power transfer. This principle has important applications in electronic circuit design. Maximum power is transferred from the power source of a circuit to the load connected to it when the load resistance (R_L) is equal to the source resistance (R_S). If R_L is less than R_S, a smaller amount of power is converted in the load. Likewise, if R_L is greater than R_S, less power is dissipated in the load.

OBJECTIVES

1. To verify that maximum power is transferred from source to load when $R_L = R_S$.

2. To use your data to construct a graph that shows the effect of changing the values of load resistance and that maximum power is developed at only one specific value of load resistance.

3. To use a multimeter to make measurements to verify the maximum power-transfer theorem.

EQUIPMENT

Variable dc power supply or 6 V battery

Multimeter (VOM)

Resistors: 15 Ω, 22 Ω, 100 Ω, 220 Ω

Potentiometer: 200 Ω

PROCEDURE

1. Calculate the Thevinin equivalent resistance (R_{TH}) of the circuit shown in Fig. 3-12A. This value is the source resistance of the circuit. $R_{TH} =$ —————— Ω.

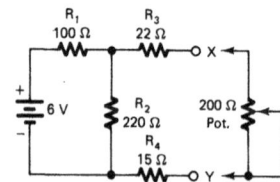

FIGURE 3-12A Circuit for calculating power.

2. Construct the circuit of Fig. 3-12A. Before applying power, connect a shorting wire across the point of the circuit where the power supply is to be connected to measure R_{TH}.

3. Use a meter to measure the resistance across the load terminals (X and Y). $R_{TH} =$ —————— Ω.

4. Remove the shorting wire and then apply 6 V to the circuit.

5. Connect a 200 Ω potentiometer as shown in Fig. 3-12A. Adjust its resistance to 20 Ω. Connect the pot across the load terminals of the circuit.

6. Measure the voltage across the load terminals with 20 Ω of R_L. Record this value in Fig. 3-12B.

7. Remove the potentiometer from the circuit and adjust its resistance to 40 Ω. Connect this resistance across the load terminals.

Load resistance (ohms)	Output voltage (volts)	Calculated power = V²/R (milliwatts)
20		
40		
60		
80		
100		
120		
140		
160		
180		
200		

FIGURE 3-12B Maximum power-transfer values.

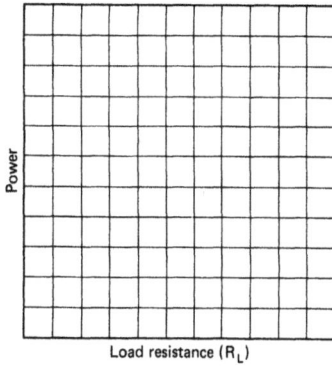

FIGURE 3-12C Graph of R_L versus power.

8. Measure and record in Fig. 3-12B the voltage across the load terminals with 40 Ω of R_L.

9. Measure the output voltage for each of the loads listed in Fig. 3-12B.

10. Calculate the power converted by each load using the formula $P = V^2/R_L$, where V is the voltage across the load (R_L). Place these values in Fig. 3-12B.

ANALYSIS

1. On the graph of Fig. 3-12C, plot the relationship of R_L (horizontal axis) and power (vertical axis) for the values of Fig. 3-12B.

2. Calculate the percentage difference between measured and calculated values of R_{TH} from steps 1 and 3. % Difference = ——— %.

3. State the maximum power-transfer theorem.

   _____

   _____

4. Why is the value of R_{TH} used as the source resistance of the circuit that you constructed?

   _____

   _____

EXPERIMENT 3-13

BRIDGE CIRCUITS

In this experiment you will study a type of electric circuit called a *bridge circuit*. Bridge circuits are used for several applications, particularly in electric measurement. A bridge circuit may be designed to measure electric component values by means of comparing an unknown value with a known or standard value. Other applications of bridge circuits include rectification circuits, which convert ac into dc.

OBJECTIVES

1. To learn to make a Thevinin equivalent circuit, which simplifies analysis of a bridge circuit.

2. To use a multimeter to make measurements for analyzing a bridge circuit.

EQUIPMENT

Multimeter (VOM)

Variable dc power supply or 6 V battery

Resistors: 470 Ω, 1 kΩ, 1.2 kΩ, 212 kΩ, 4.7 kΩ

PROCEDURE

1. Use the following procedure to calculate the value of Thevinin equivalent voltage (V_{TH}) and equivalent resistance (R_{TH}) for the bridge circuit shown in Fig. 3-13A. Label these values on the diagram.

   a. Remove the 4.7 kΩ load from the circuit.

   b. Look into the circuit from the load (X and Y) terminals to determine R_{TH}. The power supply should be replaced with a short circuit.

   c. Sketch the new circuit arrangement of the four resistors in Fig. 3-13B.

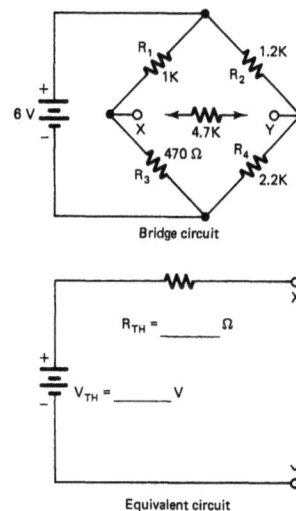

Bridge circuit

Equivalent circuit

FIGURE 3-13A Bridge circuit and equivalent circuit.

FIGURE 3-13B Sketch of a bridge circuit.

d. Calculate the R_{TH} of this arrangement.
   $R_{TH} = \underline{\qquad}\ \Omega$.

e. Disregard R_2 and R_4 and calculate the current that would flow through R_1 and R_3 if R_2 and R_4 were disconnected from the circuit. I_{R1} or I_{R3} = ———— mA with R_2 and R_4 open.

f. Disregard R_1 and R_3 and calculate the current that would flow through R_2 and R_4 if R_1 and R_3 were disconnected from the circuit. I_{R2} or I_{R4} = ———— mA.

g. Calculate the voltage drop across R_3 caused by the current determined in step e.
   V_{R4} = ———— V.

h. Calculate the voltage drop across R_4 caused by the current determined in step f.
   V_{R4} = ———— V.

i. Subtract V_{R3} from V_{R4}. This is the difference in potential (voltage drop) across points X and Y. This value is the V_{TH} of the circuit.
   V_{TH} = ———— V.

2. Construct the bridge circuit shown in Fig. 3-13A. Apply 6 V from the power supply or battery.

3. Use a meter to measure the following values with the 4.7 kΩ load connected
   a. V_{R1} = ———— V
   b. V_{R2} = ———— V
   c. V_{R3} = ———— V
   d. V_{R4} = ———— V
   e. V_{x-y} = ———— V
   f. I_{RL} = ———— mA

4. Remove the 4.7 kΩ load and make the following measurements:
   a. V_{x-y} = ———— V (V_{TH})
   b. V_{R1} = ———— V
   c. V_{R2} = ———— V
   d. V_{R3} = ———— V
   e. V_{R4} = ———— V
   f. I_{RL} = ———— mA

5. Disconnect the power supply from the circuit and connect a shorting wire across the bridge circuit where the power supply leads were connected. Measure the resistance across X and Y. This is the measured value of R_{TH}.
   R_{TH} = ———— Ω.

6. Using the Thevinin equivalent circuit values calculated in step 1, add a 4.7 kΩ resistor across the load (X and Y) terminals. Calculate the values of I_{RL} and V_{RL} with this load resistance.
   I_{RL} = ———— mA; V_{RL} = ———— V.

ANALYSIS

1. Using the Thevinin equivalent circuit values of R_{TH} and V_{TH} for the bridge circuit, calculate the load current (I_{RL}) and voltage across the load (V_{RL}) for the following values of load resistance:

   a. $R_L = 200\ \Omega$, I_{RL} = ——— mA; V_{RL} = ——— V

   b. $R_L = 1.5\ k\Omega$, I_{RL} = ——— mA; V_{RL} = ——— V

   c. $R_L = 2.7\ k\Omega$, I_{RL} = ——— mA; V_{RL} = ——— V

2. Compare the calculated and measured values of V_{TH} from steps 1i and 4a. Calculate the percentage difference of these values.
   % Difference = ——————— %.

3. What effect does the addition of a load resistance have on the voltage across R_1, R_2, R_3, and R_4? (See steps 3 and 4.)

   _____

   _____

4. Calculate the percentage difference of the measured and calculated values of I_{RL} from steps 3f and 6. % Difference = ——————— %.

Ohm's Law and Electric Circuits

Instructions: For each of the following, circle the answer that most correctly completes the statement.

1. Ohm's law, stated mathematically, could best be expressed by the formula
   a. $X = 2\pi fL$
   b. $V = I \times R$
   c. $R = I \times V$
   d. $P = R \times V$

2. In applying Kirchhoff's voltage law to a circuit, a voltage drop is labeled negative when
   a. There is a significant voltage drop
   b. There is a very small voltage drop
   c. Going from a negative to a positive polarity
   d. Going from a positive to a negative polarity

3. As voltage increases in a circuit, the power consumed
   a. Decreases
   b. Increases
   c. Remains the same
   d. Is cut off

4. A parallel circuit has
   a. One path of current flow
   b. Two or more paths of current flow
   c. The same current in all parts of the circuit
   d. At least eight components

5. A system in which algebraic equations are used to solve complex circuit problems is
   a. Ohm's law
   b. Atomic theory
   c. Kirchhoff's law
   d. Newton's law

6. A lamp has a voltage of 110 V applied and a current flow of 0.9 A. The resistance of the lamp bulb is
   a. 12.22 Ω
   b. 122.2 Ω
   c. 0.008 Ω
   d. 0.08 Ω

7. The total resistance of the circuit shown in Fig. E-7 is
   a. 40 Ω
   b. 60 Ω
   c. 5.45 Ω
   d. 11/60 Ω

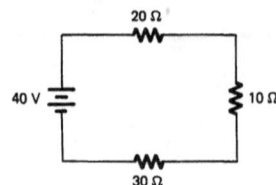

FIGURE 3E-7

8. The total current of the circuit shown in Fig. E-7 is
   a. 6 A
   b. 0.166 A
   c. 0.67 A
   d. 1.66 A

9. The total resistance of the circuit shown in Fig. E-9 is
   a. 7/40 Ω
   b. 0.176 Ω
   c. 5.71 Ω
   d. 70 Ω

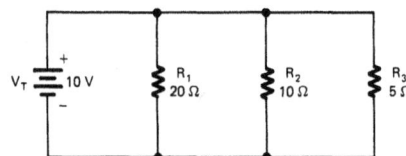

FIGURE 3E-9

10. The power consumed by the 10 Ω resistor in the circuit shown in Fig. E-9 is
    a. 10 W
    b. 20 W
    c. 30 W
    d. 40 W

11. The total current flow in the circuit shown in Fig E-11 is
    a. 60 A
    b. 60 μA
    c. 6 A
    d. 600 mA

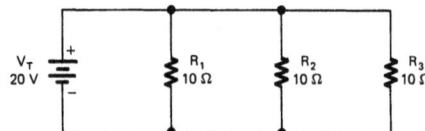

FIGURE 3E-11

12. The current flow through the 40 Ω resistor in the circuit of Fig. E-12 is
    a. 0.56 A
    b. 0.221 A
    c. 0.144 A
    d. 0.360 A

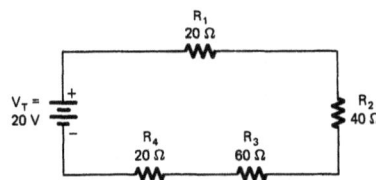

FIGURE 3E-12

13. The total resistance of the circuit shown in Fig. E-13 is
    a. 1.33 Ω
    b. 0.75 Ω
    c. 14 Ω
    d. 9.33 Ω

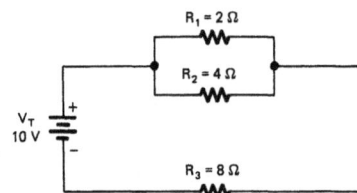

FIGURE 3E-13

14. The total power consumed by the circuit shown in Fig. E-13 is
    a. 93.3 W
    b. 140 W
    c. 75 W
    d. 133 W

15. A lamp has 110 V applied and a current flow of 0.9 A is produced. What is the resistance of the lamp?
    a. 12.22 Ω
    b. 122.2 Ω
    c. 0.008 Ω
    d. 0.08 Ω

16. As current increases in an electric circuit, power consumed
    a. Decreases
    b. Increases
    c. Remains the same
    d. Drops to zero

17. A series circuit has
    a. One path of current flow
    b. At least four paths for current flow
    c. Two or more paths for current flow
    d. The same current in all parts of the circuit

18. A system in which algebraic equations are not used for solving circuit problems is
    a. Ohm's law
    b. Atomic law
    c. Kirchhoff's law
    d. Newton's law

19. The voltage across each component is the same in a
    a. Series circuit
    b. Combination circuit
    c. Parallel circuit
    d. Resistive circuit

20. Current is related to the resistance and voltage in an electric circuit as follows:
   a. Current is directly proportional to resistance and inversely proportional to voltage.
   b. Current is inversely proportional to resistance and directly proportional to voltage.
   c. Current is inversely proportional to both resistance and voltage.
   d. Current is directly proportional to both resistance and voltage.

True-False: Place either T or F in each blank.

_____ 21. The current in a parallel circuit is the same through all parts of the circuit.

_____ 22. The voltage in a series circuit is the same across all circuit components.

_____ 23. The voltage drop across an electric component may be found by means of multiplying the current flowing through that component by its resistance value.

_____ 24. A dc electric circuit is one in which the current path is from a negative potential to a positive potential.

_____ 25. A resistor with a 1 W power rating would be able to safely conduct more electric current than a 5 W resistor.

_____ 26. Kirchhoff's law is a problem-solving method that may be applied to electric circuits that have more than one voltage source.

_____ 27. Essentially, Kirchhoff's voltage law states that the sum of voltage drops in an electric circuit will be somewhat less than the source voltage.

_____ 28. The total resistance in a parallel circuit is always less than the smallest individual resistance.

_____ 29. We may replace a complex linear network with a voltage source and equivalent resistance by applying Thevinin's theorem.

_____ 30. Norton's theorem states that a circuit containing only linear bilateral elements can be reduced to as many individual circuits as there are power sources.

Magnetism and Electromagnetism

Magnetism has been studied for many years. Some metals in their natural state attract small pieces of iron. This property is called *magnetism.* Materials that have this ability are called *natural magnets.* The first magnets used were called *lodestones.* Now artificial magnets are made in many different strengths, sizes, and shapes. Magnetism is important because it is used in electric motors, generators, transformers, relays, and many other electric devices. The earth itself has a magnetic field like a large magnet.

Electromagnetism is magnetism that is brought about by electric current flow. Many electric machines operate because of electromagnetism. This unit deals with magnetism, electromagnetism, and some important applications.

UNIT OBJECTIVES

Upon completion of this unit, you will be able to do the following:

1. Define the various terms relative to magnetism.
2. Explain the operation of various magnetic devices.
3. State Faraday's law for electromagnetic induction.
4. List three factors that affect the strength of electromagnets.
5. Apply the left-hand rule for polarity.
6. Describe the construction of a relay and solenoid.
7. Define the terms *residual magnetism, permeability, retentivity, magnetic saturation,* and *magnetizing force.*
8. Describe the domain theory of magnetism.

Important Terms

In this unit you will study magnetism and electromagnetism. There are many important applications of magnetism and electromagnetism. The terms that follow will help you define some of the specifics involved in the study of magnetism and electromagnetism.

Alnico An alloy of aluminum, nickel, iron, and cobalt used to make permanent magnets.

Ampere-turn The unit of measurement of magnetic field strength; amperes of current times the number of turns of wire.

Armature The movable part of an electromagnetic relay.

Coefficient of coupling (k) A decimal value that indicates the amount of magnetic coupling between coils.

Core Iron or steel materials used to make laminated internal sections of electromagnets.

Core saturation The condition in which the atoms of a metal core material are aligned in the same pattern so that no more magnetic lines of force can be developed.

Coupling The amount of mutual inductance between coils.

Domain theory A theory of magnetism in which it is assumed that groups of atoms produced by movement of electrons align themselves in groups called *domains* in magnetic materials.

Electromagnet A coil of wire wound on an iron core so that it becomes magnetized as current flows through the coil.

Flux See *magnetic flux.*

Flux density The number of lines of force per unit area of a magnetic material or circuit.

Gauss A unit of measurement of magnetic flux density.

Gilbert A unit of measurement of magnetomotive force (MMF).

Hysteresis The property of a magnetic material that causes actual magnetizing action to lag behind the force that produces it.

Laws of magnetism (1) Like magnetic poles repel. (2) Unlike magnetic poles attract.

Lines of force Same as *magnetic flux.*

Lodestone The name used for natural magnets used in early times.

Magnet A metallic material, usually iron, nickel, or cobalt, that has magnetic properties.

Magnetic circuit A complete path for magnetic lines of force from a north to a south polarity.

Magnetic field Magnetic lines of force that extend from a north polarity and enter a south polarity to form a closed loop around the outside of a magnetic material.

Magnetic flux (Φ) Invisible lines of force that extend around a magnetic material.

Magnetic materials Metallic materials, such as iron, nickel, and cobalt, that exhibit magnetic properties.

Magnetic poles Areas of concentrated lines of force on a magnet that produce north and south polarities.

Magnetic saturation A condition that exists in a magnetic material when an increase in magnetizing force does not produce an increase in magnetic flux density around the material.

Magnetomotive force (MMF) A force that produces magnetic flux around a magnetic device.

Magnetostriction The effect that produces a change in shape of certain materials when they are placed in a magnetic field.

Natural magnet Metallic materials that have magnetic properties in their natural state.

Permanent magnet Bars or other shapes of materials that retain their magnetic properties.

Permeability (μ) The ability of a material to conduct magnetic lines of force as compared with air; the ability of a material to magnetize or demagnetize.

Polarities See *magnetic poles.*

Relay An electromagnetic switch that uses a low current to control a higher-current circuit.

Reluctance (ℜ) The opposition of a material to the flow of magnetic flux.

Residual magnetism The magnetism that remains around a material after the magnetizing force has been removed.

Retentivity The ability of a material to retain magnetism after a magnetizing force has been removed.

Solenoid An electromagnetic coil with a metal core that moves when current passes through the coil.

Permanent Magnets

Magnets are made of iron, cobalt, or nickel, usually in an alloy combination. An alloy is a mixture of these materials. Each end of the magnet is called a *pole.* If a magnet were broken, each part would become a magnet. Each magnet would have two poles. Magnetic poles are always in pairs. When a magnet is suspended in air so that it can turn freely, one pole will point to the North Pole. The earth is like a large permanent magnet. This is why compasses can be used to determine direction. The north pole of a magnet *attracts* the south pole of another magnet. A north pole *repels* another north pole and a south pole *repels* another south pole. The two laws of magnetism are (1) like poles repel and (2) unlike poles attract.

The magnetic field patterns that exist when two permanent magnets are placed end to end are shown in Fig. 4-1. When the magnets are farther apart, a smaller force of attraction or repulsion exists. This type of permanent magnet is called a *bar magnet*.

Some materials retain magnetism longer than others do. Hard steel holds its magnetism much longer than soft steel does. A *magnetic field* is set up around any magnetic material. The field is made up of *lines of force,* or *magnetic flux.* These magnetic flux lines are invisible. They never cross one another, but they always form individual closed loops around a magnetic material. They have a definite direction from north to south pole along the outside of a magnet. When magnetic flux lines are close together, the magnetic field is strong. When magnetic flux lines are farther apart, the field is weaker. The magnetic field is strongest near the

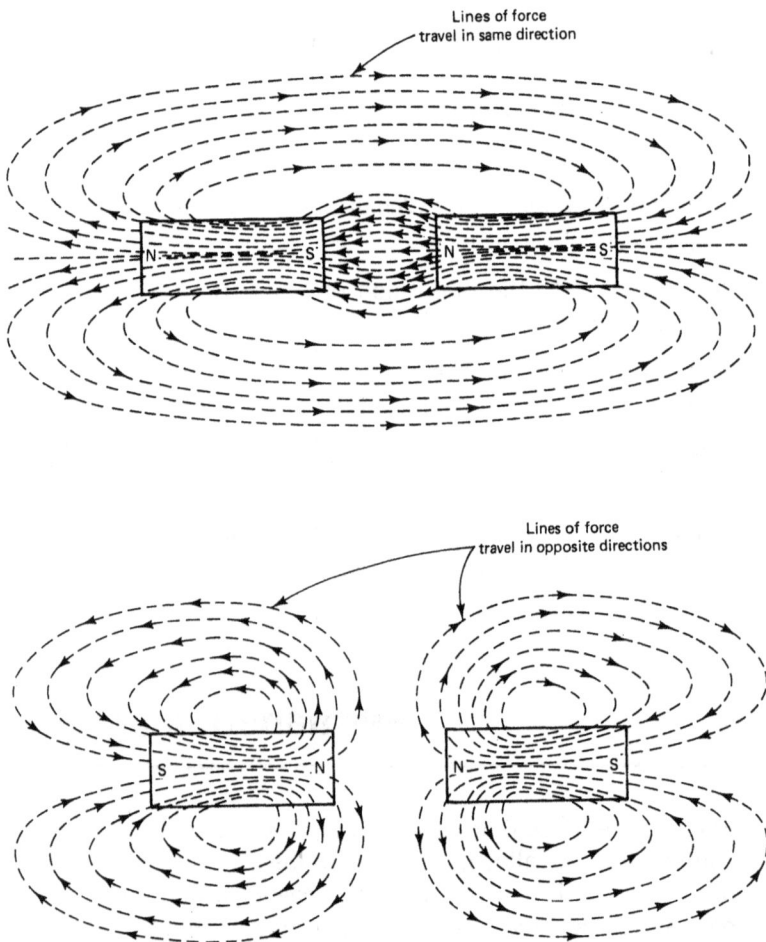

FIGURE 4-1 Magnetic field patterns that exist when magnets are placed end to end.

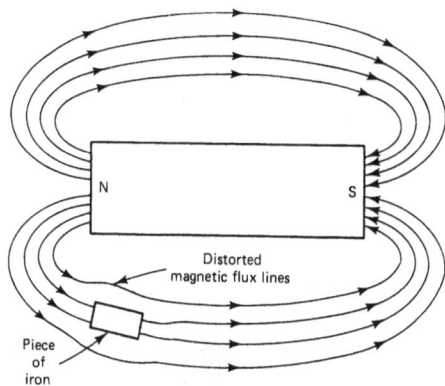

FIGURE 4-2 Magnetic flux lines distorted while passing through a piece of iron.

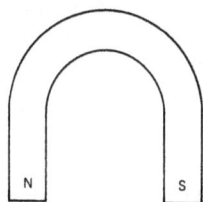

FIGURE 4-3 Horseshoe magnet.

poles. Lines of force pass through all materials. It is easy for lines of force to pass through iron and steel. Magnetic flux passes through a piece of iron as shown in Fig. 4-2.

When magnetic flux passes through a piece of iron, the iron acts as a magnet. Magnetic poles are formed because of the influence of the flux lines. These are called *induced poles*. The induced poles and the poles of the magnet attract and repel each other. Magnets attract pieces of soft iron this way. It is possible to magnetize pieces of metal temporarily by using a bar magnet. If a magnet is passed over the top of a piece of iron several times in the same direction, the soft iron becomes magnetized. It stays magnetized for a short time.

When a compass is brought near the north pole of a magnet, the north-seeking pole of the compass is attracted to it. The polarities of the magnet can be determined by means of observing a compass brought near each pole. Compasses detect the presence of magnetic fields.

Horseshoe magnets are similar to bar magnets. They are bent in the shape of a horseshoe, as shown in Fig. 4-3. This shape gives more magnetic field strength than a similar bar magnet, because the magnetic poles are closer. The magnetic field strength is concentrated in one area. Many electric devices use horseshoe magnets.

A magnetic material can lose some of its magnetism if it is jarred or heated. People must be careful when handling equipment that contains permanent magnets. A magnet also becomes weakened by loss of magnetic flux. Magnets should always be stored with a *keeper*, which is a soft-iron piece used to join magnetic poles. The keeper provides the magnetic flux with an easy path between poles. The magnet retains its greatest strength for a longer period of time if keepers are used. Bar magnets should always be stored in pairs with a north pole and a south pole placed together. A complete path for magnetic flux is made in this way.

Magnetic Field Around Conductors

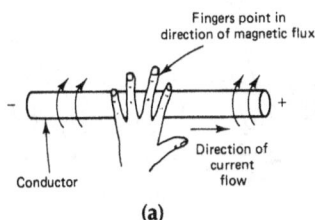

FIGURE 4-4 (a) Left-hand rule of magnetic flux.

Current-carrying conductors produce a magnetic field. It is possible to show the presence of a magnetic field around a current-carrying conductor. A compass is used to show that the magnetic flux lines are circular. The conductor is in the center of the circle. The direction of the current flow and the magnetic flux lines can be shown with the *left-hand rule* of magnetic flux. A conductor is held in the left hand, as shown in Fig. 4-4a. The thumb points in the direction of current flow from negative to positive. The fingers then encircle the conductor in the direction of the magnetic flux lines.

A circular magnetic field is produced around a conductor. The field is stronger near the conductor and becomes weaker farther away from the conductor. A cross-sectional end view of a conductor with current flowing toward the observer is shown in Fig. 4-4b. Current flow toward the observer is shown by a circle with a dot in the center. The direction of the magnetic flux lines is clockwise. This can be verified with the left-hand rule.

When the direction of current flow through a conductor is reversed, the direction of the magnetic lines of force also is reversed. The cross-sectional end view of a conductor in Fig. 4-4c shows a current flow in a direction away from the observer. The direction of the magnetic lines of force is counterclockwise.

The presence of magnetic lines of force around a current-carrying conductor can be observed with a compass. When a compass is moved around the outside of a conductor, the needle aligns itself tangentially to the lines of force, as shown in Fig. 4-4d. The needle does not point toward the conductor. When current flows in the opposite direction, the compass polarities reverse. The compass needle aligns itself tangentially to the conductor.

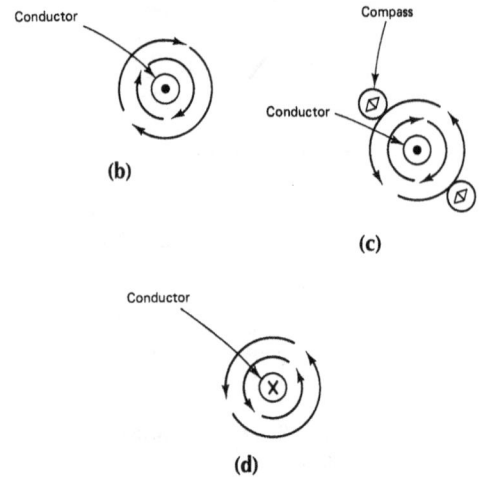

(b)

(c)

(d)

FIGURE 4-4 (b) Cross section of conductor with current flow toward observer. (c) Compass aligns tangent to magnetic lines of force. (d) Cross section of conductor with current flow away from the observer.

Magnetic Field Around a Coil

The magnetic field around one loop of wire is shown in Fig. 4-5. Magnetic flux lines extend around the conductor as shown. Inside the loop, the magnetic flux is in one direction. When many loops are joined together to form a coil, the magnetic flux lines surround the coil as shown in Fig. 4-6. The field around a coil is much stronger than the field of one loop of wire. The field around the coil is the same shape as the field around a bar magnet. A coil that has an iron or steel core inside it is called an electromagnet. A core increases the magnetic flux density of a coil.

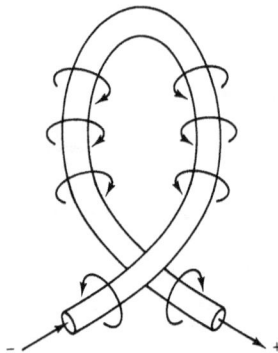

(a)

(b)

(c)

FIGURE 4-6 Magnetic field around a coil. (a) Coil of wire showing current flow. (b) Lines of force around two loops that are parallel. (c) Cross section of coil showing lines of force.

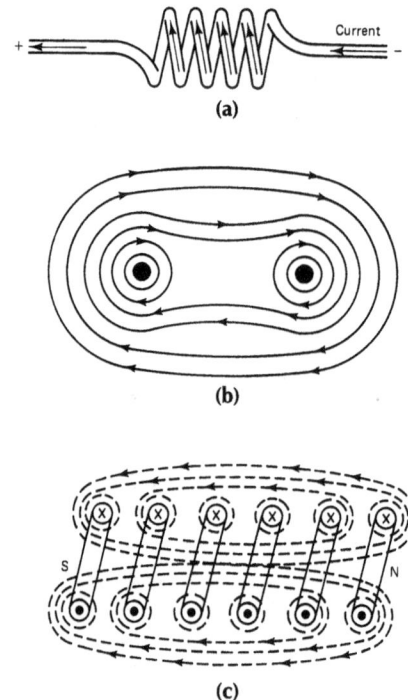

FIGURE 4-5 Magnetic field around a loop of wire.

Self-Examination

1. A substance that has its molecules arranged at random is _____.

2. A substance that has its molecules aligned is _____.

3. The space surrounding a magnet, in which the magnetic force acts, is called the _____.

4. Magnetic lines of force are directed from the _____ pole to the _____ pole.

5. Like magnetic poles _____, unlike poles _____.

Answers

1. Nonmagnetic	2. Magnetic
3. Magnetic field	4. North, south
5. Repel, attract	

Electromagnets

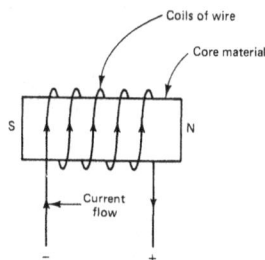

FIGURE 4-7 An electromagnet.

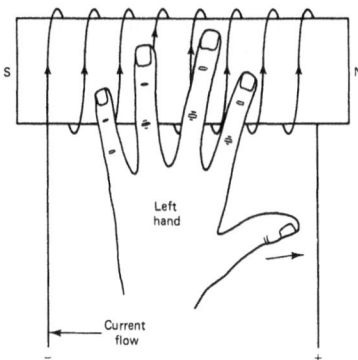

FIGURE 4-8 Left-hand rule for finding the polarities of an electromagnet.

An electromagnet is produced when current flows through a coil of wire, as shown in Fig. 4-7. The north pole of a coil of wire is the end where the lines of force come out. The south polarity is the end where the lines of force enter the coil. This is like the field of a bar magnet. To find the north pole of a coil, use the *left-hand rule for polarity,* as shown in Fig. 4-8. Grasp the coil with the left hand. Point the fingers in the direction of current flow through the coil. The thumb points to the north polarity of the coil.

When the polarity of the voltage source is reversed, the magnetic poles of the coil also is reversed. The poles of an electromagnet can be checked with a compass. The compass is placed near a pole of the electromagnet. If the north-seeking pole of the compass points to the coil, that side is the north pole.

Electromagnets have several turns of wire around a soft iron core. An electric power source is then connected to the ends of the turns of wire. When current flows through the wire, magnetic polarities are produced at the ends of the soft iron core. The three basic parts of an electromagnet are (1) an iron core, (2) wire windings, and (3) an electric power source. Electromagnetism is made possible by electric current flow, which produces a magnetic field. When electric current flows through the coil, the properties of magnetic materials are developed.

Magnetic Strength of Electromagnets

The magnetic strength of an electromagnet depends on the three following factors: (1) the amount of current passing through the coil, (2) the number of turns of wire, and (3) the type of core material. The number of magnetic lines of force is increased by means of increasing the current, increasing the number of turns of wire, or using a more desirable type of core material. The magnetic strength of electromagnets is determined by the *ampere-turns* of each coil. The number of ampere-turns is equal to the current in amperes multiplied by the number of turns of wire ($I \times N$). For example, 200 ampere-turns is produced by 2 A of current through a 100-turn coil. One ampere of current through a 200-turn coil produces the same magnetic field strength. Figure 4-9 shows how the magnetic field strength of an electromagnet changes with the number of ampere-turns.

The magnetic field strength of an electromagnet also depends on the type of core material. Cores usually are made of soft iron or steel. These materials transfer a magnetic field better than air or other nonmagnetic materials. Iron cores increase the *flux density* of an electromagnet. Figure 4-10 shows that an iron core causes the magnetic flux to be denser.

An electromagnet loses its field strength when the current stops flowing. However, the core of an electromagnet retains a small amount of magnetic strength after current stops flowing. This is called *residual magnetism,* or "leftover" magnetism. It can be reduced by using soft iron cores or increased by hard steel core material. Residual magnetism is important in the operation of some types of electric generators.

In many ways electromagnetism is similar to magnetism produced by natural magnets such as bar magnets. However, the main advantage of electromagnetism is that it is easily controlled. It is easy to increase the strength of an electromagnet by means of increasing the current flow through the coil. This is done by increasing the voltage applied to the coil. The second way to increase the strength of an electromagnet is to have more turns of wire around the core. A greater number of turns produces more magnetic lines of force around the electromagnet. The strength of an electromagnet also is affected by the type of core material used. Different alloys of iron are used to make the cores of electromagnets. Some materials aid in the development of magnetic lines of force to a greater extent. Other types of core materials offer greater resistance to the development of magnetic flux around an electromagnet.

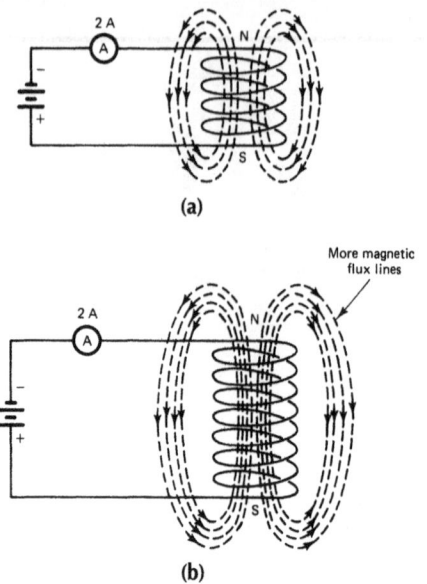

FIGURE 4-9 Effect of ampere turns on magnetic field strength. (a) Five turns, two amperes = 10 ampere-turns. (b) Eight turns, two amperes = 16 ampere-turns.

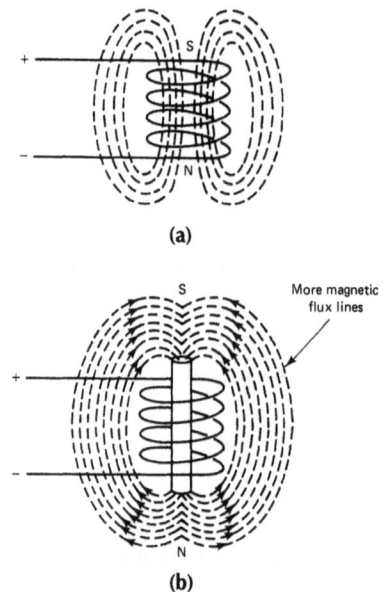

FIGURE 4-10 Effect of iron core on magnetic strength. (a) Coil without core. (b) Coil with core.

Ohm's Law for Magnetic Circuits

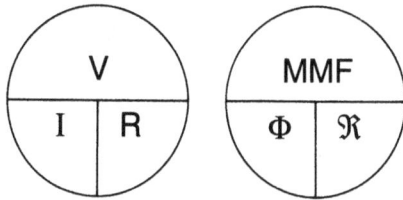

FIGURE 4-11 Relations of magnetic and electric terms.

Ohm's law for electric circuits is described in unit 2. A similar relation exists in magnetic circuits. Magnetic circuits have *magnetomotive force* (MMF), magnetic flux (Φ), and reluctance ($\Re$). MMF is the force that causes magnetic flux to be developed. *Magnetic flux* describes the lines of force around a magnetic material. *Reluctance* is opposition to the development of a magnetic flux. These terms may be compared with voltage, current, and resistance in electric circuits, as shown in Fig. 4-11. When MMF increases, magnetic flux increases. In an electric circuit, when voltage increases, current increases. When resistance in an electric circuit increases, current decreases. When reluctance of a magnetic circuit increases, magnetic flux decreases. Study the relation of magnetic and electric terms in Fig. 4-11.

Domain Theory of Magnetism

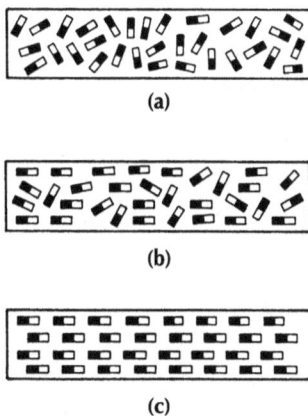

FIGURE 4-12 Domain theory of magnetism.
(a) Unmagnetized. (b) Slightly magnetized.
(c) Fully magnetized saturation.

A theory of magnetism was presented in the nineteenth century by a German scientist named Wilhelm Weber. Weber's theory of magnetism was called the *molecular theory*. It dealt with alignment of molecules in magnetic materials. Weber believed that molecules were aligned in an *orderly* arrangement in magnetic materials. He thought that in nonmagnetic materials, molecules were arranged in a *random* pattern.

Weber's theory has been modified to become the *domain theory* of magnetism. This theory deals with alignment of domains in materials rather than molecules. A domain is a group of atoms (about 10^{15} atoms). Each domain acts like a tiny magnet. The rotation of electrons around the nucleus of these atoms is important. Electrons have a negative charge. As they orbit around the nucleus of atoms, their electric charges move. This moving electric field produces a magnetic field. The polarity of the magnetic field is determined by the direction of electron rotation.

The domains of magnetic materials are atoms grouped together. Their electrons are believed to spin in the same direction. This produces a magnetic field because of electric charge movement. Figure 4-12 shows the arrangement of domains in magnetic, nonmagnetic, and partially magnetized materials. In *nonmagnetic materials,* half the electrons spin in one direction and half spin in the opposite direction. Their charges cancel each other out. No magnetic field is produced because the charges cancel. In *magnetic materials* electrons rotate in the same direction. This causes the domains to act like tiny magnets that align to produce a magnetic field.

Electricity Produced by Magnetism

A scientist named Michael Faraday discovered in the early 1830s that electricity is produced by magnetism. He found that if a magnet is placed inside a coil of wire, electric current is produced when the magnet is moved. Faraday found that electric current is produced by magnetism and motion.

Faraday's law is stated as follows: *When a coil of wire moves across the lines of force of a magnetic field, electrons flow through the wire in one direction. When the coil of wire moves across the magnetic lines of force in the opposite direction, electrons flow through the wire in the opposite direction.*

This law is the principle of electric power generation produced by magnetism. Figure 4-13 shows Faraday's law as it relates to electric power generation.

Current flows in a conductor placed inside a magnetic field only when there is motion between the conductor and the magnetic field. If a conductor is stopped while it is moving across the magnetic lines of force, current stops flowing. The operation of electric generators depends on conductors moving across a magnetic field. This principle is called *electromagnetic induction*.

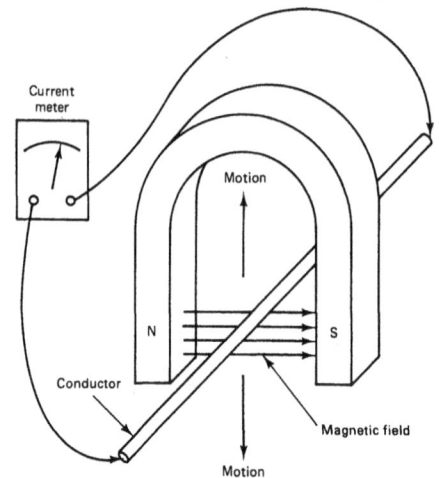

FIGURE 4-13 Faraday's law. Electric current is produced when there is relative motion between a conductor and a magnetic field.

Relays

Relays are electric devices that rely on magnetism to operate. They control other equipment, such as motors, lights, or heating elements. Relays are very important devices. They are used to start the operation of other equipment. They use a small amount of electric current to control a larger current, such as the current through a motor.

The basic construction and symbols of a relay are shown in Fig. 4-14. Relays have an electromagnetic coil with electric power applied to its two external leads. When the power is turned on, the electromagnet is energized. The electromagnet part of the relay controls a set of contacts. The contacts are called *normally open* (NO) or *normally closed* (NC), depending on their condition when the electromagnet is not energized. There is also a *common* contact.

If a lamp and its power source are connected in series with the common and NO contact as shown in Fig. 4-14a, the lamp is off when the relay is not energized. The

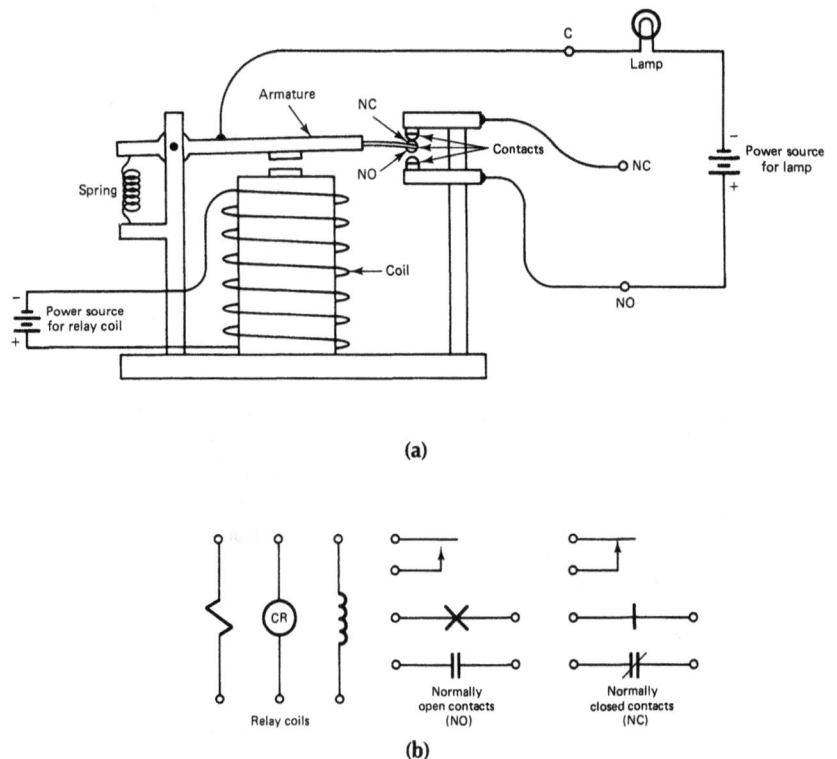

(a)

(b)

FIGURE 4-14 Construction and symbols of a relay. (a) Pictorial. (b) Symbols.

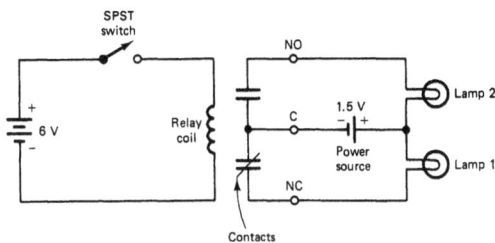

FIGURE 4-15 Schematic of a relay circuit.

lamp or any load connected to the relay contacts requires a separate power source. If the relay is energized when power is applied, the common contact is attracted to the NO contact by magnetic energy. The common contact is built onto a movable armature that moves when the electromagnet is energized. When the relay is energized, the light connected to the NO contact turns on.

The NC contacts are used in a similar way (Fig. 4-15). When the relay is off, the circuit from the common terminal through the power source and lamp 1 is closed. This causes lamp 1, which is in series with the NC contacts, to be turned off. The power source for the lamps is in the common line because it is common to both of the other contacts. When the relay is turned on, the lamps change states. Lamp 2, connected to the NO contact, is turned on, whereas lamp 1, connected to the NC contact, is turned off. In Fig. 4-14a the common contact moves from the NC contact so that it touches the NO contact. This shows the basic operation of a relay with NO and NC contacts. Such a relay is very common. It is called a double-pole, single-throw (DPST) relay. Many other types also are used.

The coil resistance of a relay is determined by the size of the wire used to wind the coil and the number of windings. A coil with only a few turns of large-diameter wire has a low resistance. This causes a high current flow through the relay coil. If a relay coil has many turns of small-diameter wire, it has a high resistance. Small wire has high resistance, and large wire has lower resistance to current flow. Coil resistance usually is marked on a relay. It also may be measured with a meter.

There are some important current ratings for relays. Two of these ratings are called *pickup current* and *dropout current*. These ratings usually are specified on the relay. They also may be found by use of a variable power source and a current meter to measure the actual values. When the voltage applied to the relay coil is increased to the point at which the relay turns on, the current indicated on the meter is the *pickup current*. Pickup current is the minimum current needed to energize the relay. If the applied voltage is decreased until the relay deenergizes, the meter indicates *dropout current*. Dropout current is the minimum current that will keep the relay energized. The *contact current* rating is very important. A large current usually flows through the relay contacts and the load connected to the relay. This rating is the maximum current that can safely flow through the contact circuit.

Solenoids

Solenoids are similar to relays because they also use electromagnetic coils. They are sometimes called *actuators*. Instead of having contacts, a solenoid has a plunger that moves when the coil is energized. The back of the plunger is attached to a spring. It causes the plunger to return to its original position when the solenoid deenergizes. The movement of the plunger of the solenoid is used to activate some type of load connected to it. When the solenoid is energized, it moves the plunger in the center. A solenoid can be used to open a control valve to allow liquid from a tank to flow into a container. When the solenoid is deenergized, the control valve closes and stops the flow of liquid.

Magnetic Motor Contactors

An important type of relay is an *electric motor contactor.* The motor contactor is a control element that starts and stops motors. It operates through electromagnetic relay action. A "start" push button switch is pressed to close a contactor. This completes a low-current path through the contactor coil. The contactor coil produces a magnetic field that causes a set of contact points to close. The movement of a part called an *armature* completes an electric path between the power line and a motor. When this action takes place, the motor starts.

Releasing the start button does not deenergize the contactor coil. A path to the voltage source coil is completed through the "stop" pushbutton switch. The motor continues to run as long as electric power is applied.

A contactor-controlled motor is stopped by means of pushing the stop button. This opens the contactor voltage source of the coil. The coil deenergizes and causes the armature to move. The contact points then break contact. This removes the power from the motor. The path opens and the motor stops. Motor contactors are designed to latch in place to hold the contactor in place once the coil is energized. This characteristic is important for motor control.

Magnetic Effects

There are several magnetic effects that are important. Among these effects are residual magnetism, permeability, retentivity, and magnetic saturation.

Residual magnetism is important is in the operation of some types of electric generators. Residual magnetism is the ability of electromagnets to retain a small magnetic field after electric current is turned off. A small magnetic field remains around an electromagnet after it is demagnetized. This magnetic field is very weak.

Permeability (μ) is the ability of a magnetic material to transfer magnetic flux. It is the ability of a material to magnetize and demagnetize. Soft iron has a high permeability. It transfers magnetic flux very easily. Soft iron magnetizes and demagnetizes rapidly. This makes soft iron a good material to use in the construction of generators, motors, transformers, and other electromagnetic machines.

A related term is *relative permeability* (μ_r). This is a comparison of the permeability of a material with the permeability of air (1.0). Suppose that a material has a relative permeability of 1000. This means that the material transfers 1000 times more magnetic flux than an equal amount of air. The highest relative permeability of materials is more than 5000.

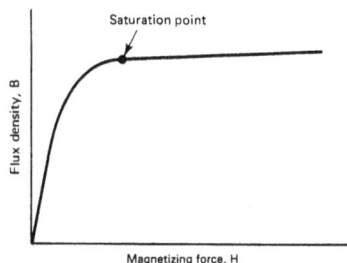

FIGURE 4-16 Magnetization or *B-H* curve.

Another magnetic effect is called *retentivity.* The retentivity of a material is its ability to retain a magnetic flux after a magnetizing force is removed. Some materials retain a magnetic flux for a long period of time. Other materials lose their magnetic flux almost immediately after the magnetizing force is removed.

Magnetic saturation is important in the operation of machines that have electromagnets, especially generators. Saturation is best explained with the curve shown in Fig. 4-16. This is called *magnetization* or *B-H* curve. The curve shows the relation between a magnetizing force (*H*) and flux density (*B*). As a magnetizing force increases, so does flux density. Flux density is the number of lines of force per unit area of a material. An increase in flux density occurs until magnetic saturation is reached. This saturation point depends on the type of material. At the saturation point, the maximum alignment of domains takes place in the material.

Magnetizing force (H) is measured in *oersteds.* The basic unit is the number of ampere-turns per meter of length. *Flux density (B)* is the amount of magnetic flux per unit area. The unit of flux density is the gauss per square centimeter of area.

Self-Examination

6. _____ is magnetism caused by current flow.

7. The polarity of an electromagnet may be determined by means of applying the _____.

8. The three basic parts of an electromagnet are _____, _____, and _____.

9. Three ways to increase the strength of an electromagnet are _____, _____, and _____.

10. _____ is the ease with which a material conducts magnetic lines of force.

Answers

6. Electromagnetism	7. Left-hand rule
8. Core, windings, voltage (current) source	9. More turns (windings), more current flow (voltage), better core material
10. Permeability	

EXPERIMENT 4-1

THE NATURE OF MAGNETISM

Magnetism is one of the longest-known natural forces. It was first discovered and used in ancient cultures as a curiosity. Many believed that this force was magic and therefore to be feared. The first magnets used were natural magnets called *lodestones* and were first put to practical use in navigation. Someone discovered that when these devices were suspended by a string and allowed to move freely, they would always align themselves to point to the north. Thus natural magnetism was first used for compasses.

Much later it was discovered that magnetism could be used to produce an electric current and that an electric current could be used to produce a magnetic field. This relation makes a knowledge of magnetism extremely important.

OBJECTIVE

To examine the characteristics of natural magnetism.

EQUIPMENT

Permanent magnet

Magnetic compass

PROCEDURE

1. Place a permanent magnet on the surface in front of you.

2. Examine the needle of the compass. You will find that the north pole of the compass needle (the one that points north) is painted, or otherwise colored, to differentiate it from the south pole.

3. Use the compass to identify the north pole and south pole of the permanent magnet. (Remember like poles repel; unlike poles attract.) Mark the north and south poles of your permanent magnet.

4. If the north pole of one magnet is brought near the north pole of another magnet, what happens and why? _____

5. If the south pole of one magnet is brought near the south pole of another magnet, what happens and why? _____

6. If the north pole of one magnet is brought near the south pole of another magnet, what happens and why? _____

ANALYSIS

1. Describe the reaction of like poles and unlike poles of magnetic fields.

    _____

    _____

2. Why can a compass be used to detect the presence of a magnetic field?

    _____

    _____

3. Define the following terms:
    a. Flux _____
    b. Magnetic field _____
    c. Magnetic lines of force _____

4. In what direction does magnetic flux travel both externally and internally around a magnet?

    _____

    _____

5. Sketch the magnetic field of a permanent magnet in the space provided.

EXPERIMENT 4-2

ELECTROMAGNETIC RELAYS

Relays are electromagnetic switches and are excellent examples of how a magnetic field attracts a magnetic material. These devices contain a coil that produces an electromagnetic field, an armature constructed of a magnetic material attracted by the coil, and a number of contacts or switches that open or close when the magnetic field attracts the armature.

OBJECTIVE

To study the electromagnetic characteristics of a relay.

EQUIPMENT

Multimeter (VOM)

Multicontact relay

6 V lamp with socket

Variable dc power supply

Resistor: 1 kΩ

6 V battery

Connecting wires

PROCEDURE

1. Prepare the multimeter to measure resistance. Measure and record the resistance of the relay coil. _____ Ω

2. Using the ohmmeter, determine how many NO and NC contacts are used with your relay.

   Number of NO contacts: _____

   Number of NC contacts: _____

3. Construct the circuit illustrated in Fig. 4-2A. Be sure that the variable dc power supply is adjusted to zero. The multimeter should be adjusted to measure dc current on the highest range.

4. Slowly adjust the variable dc power supply from zero until the 6 V lamp is turned on. Record the current measured with the multimeter when the relay is energized. This is the pickup current: Pickup current = _____ mA.

5. Slowly adjust the variable dc power supply toward zero until the 6 V lamp is turned off. Record the current measured with the multimeter when the relay is deenergized. This is the dropout current. Dropout current = _____ mA.

6. Turn off the variable power supply.

7. Alter the circuit so that it looks like that in Fig. 4-2B.

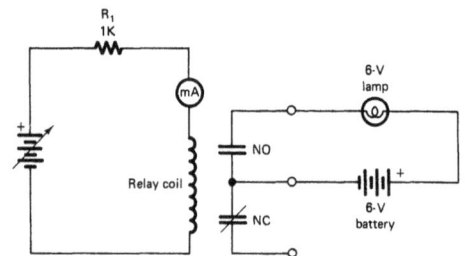

FIGURE 4-2A Relay circuit controlling a lamp.

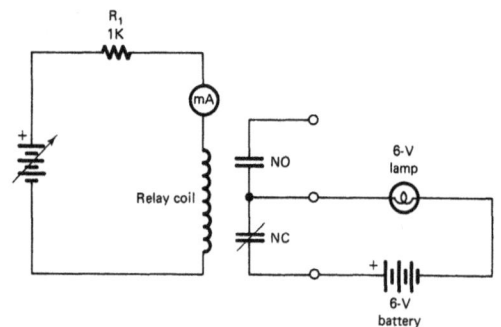

FIGURE 4-2B Modified relay circuit controlling lamp.

8. The only difference in the two circuits is the type of contacts used. In step 3 the NO contacts are used. In this procedure the NC contacts are used, causing the lamp to remain on until the relay is energized.

9. Adjust the variable dc power supply and record the pickup and dropout currents as you did in steps 4 and 5.

   Pickup I = _____ mA

   Dropout I = _____ mA

10. How do the currents recorded in step 9 compare with the current recorded in steps 4 and 5?

   _____

11. How does the action of the 6 V lamp in steps 4 and 5 compare with the action of the lamp in step 9?

   _____

ANALYSIS

1. What are normally open contacts?

   _____

2. What are normally closed contacts?

   _____

3. What is meant by the term *pickup current?*

   _____

4. What is meant by the term *dropout current?*

   _____

5. Using Ohm's law, compute the voltage across the relay coil when the relay is energized (see steps 1 and 4). $V = I \times R$ = _____ V

Unit 4 Examination

Magnetism and Electromagnetism

Instructions: For each of the following, circle the answer that most correctly completes the statement.

1. Soft iron is most suitable for use as a(n)
   a. Permanent magnet
   b. Natural magnet
   c. Electromagnet
   d. Magneto

2. Magnetic lines of force travel on the outside of a material from
   a. South to north
   b. Negative to positive
   c. Positive to negative
   d. North to south

3. A magnetic unit that corresponds to voltage is
   a. Magnetomotive force
   b. Permeability
   c. Reluctance
   d. Magnetic flux

4. In a magnetized material, the domains
   a. Are scattered in a random manner
   b. Remain in their original position
   c. Are aligned within the material
   d. Are permanently magnetized

5. A low-voltage device used to control a circuit with a higher voltage is known as
   a. A potentiometer
   b. A relay
   c. A load
   d. A diode

6. The left-hand rule of magnetism for a current-carrying conductor refers to
   a. Determining the flux density of the magnetic field that surrounds a conductor
   b. Pointing the finger of the left hand in the direction of current flow so that the thumb indicates the direction of the magnetic field

c. Determining the amount of current flow by means of an indirect approximation using the left hand

d. Pointing the thumb of the left hand in the direction of current flow so that the fingers of that hand will indicate the direction of magnetic flux

7. Which of the following magnetic terms are related, respectively, to the electric terms (1) magnetomotive force, (2) reluctance, (3) magnetic flux

a. (1) voltage, (2) current, (3) voltage

b. (1) resistance, (2) current, (3) voltage

c. (1) current, (2) resistance, (3) voltage

d. (1) voltage, (2) resistance, (3) current

8. Residual magnetism is a property of a coil or electromagnet that refers to

a. Its ability to become magnetized many times and not be destroyed

b. The excess magnetic energy produced by field coils that actually is not needed

c. Its ability to retain some magnetism for a period of time

d. The strong magnetic field that can be set up by using large field pole pieces

9. The core material for this electromagnet must have high

a. Resistance

b. Retentivity

c. Conductivity

d. Permeability

10. Using the left-hand rule, determine which end of the electromagnet shown in Fig. 4E-10 is the south pole. Also, indicate whether this same end would attract or repel the north pole of a bar magnet.

a. End A is the south pole and it will attract a north pole.

b. End B is the south pole and it will attract a north pole.

c. End A is the south pole and it will repel a north pole.

d. End B is the south pole and it will repel a north pole.

11. The magnetic field strength of the coil shown in Fig. 4E-10 can be most readily increased by

a. Increasing the number of turns on the core

b. Decreasing the permeability of the core

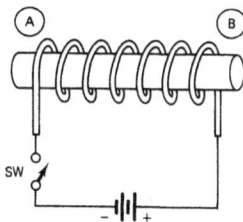

FIGURE 4E-10

c. Increasing the length of the core

d. Decreasing the applied voltage

True-False: Place either T or F in each blank.

_____ 12. When the north polarities of two permanent magnets are brought together, they attract each other.

_____ 13. Alnico is a material used for many permanent magnets.

_____ 14. A type of magnet the strength of which is determined by the number of turns of wire it has and the amount of electric current passing through these turns of wire is known as an electromagnet.

_____ 15. A magnetization curve shows the relation between a magnetizing force applied to a material and the amount of magnetic flux produced.

_____ 16. Increasing current flow through a coil increases the magnetic field around the coil.

_____ 17. In the left-hand rule of magnetism, the left thumb points to the south polarity of an electromagnetic coil.

_____ 18. Soft iron has a high permeability.

_____ 19. Three naturally magnetic materials are iron, nickel, and silver.

_____ 20. Permeability is the ability of a material to retain magnetism.

Electronic Instruments

Electronic instruments of many types are in use today. They measure many different quantities. Instruments are used to measure electric quantities and other physical quantities. In unit 2, you learned to use a multimeter (VOM) to measure voltage, current, and resistance. These are basic electric quantities.

All instruments have common characteristics. A quantity is monitored either periodically or continuously. A visual display of the quantity must be presented. Several types of instruments are used for measuring electric quantities. The basic types of instruments may be classified as (1) analog instruments, (2) comparison instruments, (3) cathode-ray tube (CRT) instruments, (4) numerical readout instruments, and (5) chart-recording instruments.

UNIT OBJECTIVES

Upon completion of this unit you will be able to do the following:

1. Calculate the value of shunt resistance needed to increase the current capability of a meter movement.

2. Calculate the value of series multiplier resistance needed to increase the full-scale voltage capability of a meter movement.

3. Describe analog meter movement.

4. Calculate the sensitivity of a voltmeter.

5. Explain the loading effect of voltmeters.

6. Draw a simple schematic of a current meter, voltmeter, or ohmmeter circuit.

Important Terms

Review the following terms before studying electronic instruments in this unit.

Ammeter A meter used to measure current flow.

Bridge circuit A circuit with groups of components that are connected by a bridge in the center and used for precision measurements.

Cathode-ray tube (CRT) A large vacuum tube in which electrons emitted from a cathode are formed into a narrow beam and accelerated and on striking a phosphorescent screen produce a visible pattern of light energy.

d'Arsonval meter movement The internal portion of an analog meter, which has a stationary permanent magnet, an electromagnetic coil, and a pointer that moves in direct proportion to the current flow through the coil.

Galvanometer A meter used to measure very small current values.

Loading The effect caused when a meter is connected into a circuit that causes the meter to draw current from the circuit.

Megohmmeter A meter used to measure very high resistances; also called a megger.

Moving coil *See d'Arsonval meter movement.*

Multiplier A resistance connected in series with a meter movement to extend the range of voltage measured.

Ohmmeter A meter used to measure resistance.

Ohms per volt rating (Ω/V) The rating that indicates the loading effect of a meter on a circuit when making a measurement.

Oscilloscope An instrument that has a cathode-ray tube to allow visual display of voltages.

Sensitivity *See ohms per volt rating.*

Shunt A resistance connected in parallel with a meter movement to extend the range of current measured.

Voltmeter A meter used to measure voltage.

Watt-hour A unit of energy measurement equal to 1 watt per hour.

Watt-hour meter A meter used to measure the rate of power conversion.

Wattmeter A meter used to measure power conversion.

Wheatstone bridge A bridge circuit used to make precision resistance measurements by means of comparing an unknown resistance with a known or standard resistance value.

X axis A horizontal line on a standard graph.

Y axis A vertical line on a standard graph.

Instruments that rely on the motion of a hand or pointer are called *analog instruments*. A volt-ohm-milliammeter (VOM), an instrument used for measuring several electric quantities, is one type of hand-deflection instrument. Single-function meters also are used to measure electric quantities. They measure only one quantity.

The basic part of an analog meter is called a meter movement. Physical quantities such as airflow or fluid pressure can be measured with analog meters. The movement of the hand or pointer over a calibrated scale indicates the quantity being measured.

Many analog meters use the *d'Arsonval, or moving-coil*, type of meter movement. The construction details of this meter movement are shown in Fig. 5-1. The hand or pointer of the movement stays on the left side of the calibrated scale. A moving coil is located inside a horseshoe magnet. Current flows through the coil from the circuit being tested. A reaction occurs between the electromagnetic field and the horseshoe magnet. This reaction causes the hand to move toward the right side of the scale. This moving-coil meter movement operates on the same principle as an electric motor. It can be used for single-function meters that measure only one quantity. It can also be used for multifunction meters, such as VOMs or digital voltmeters (DVMs), which are used to measure more than one quantity. D'Arsonval meter movements can be used to measure voltage, current, or resistance. Resistors of proper value are connected internally to the meter movement for making these measurements.

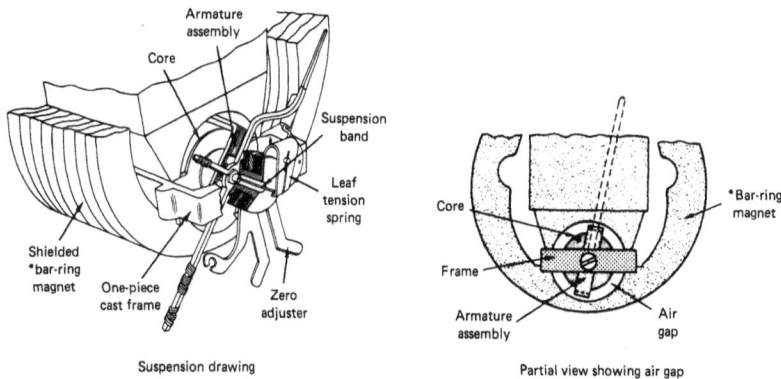

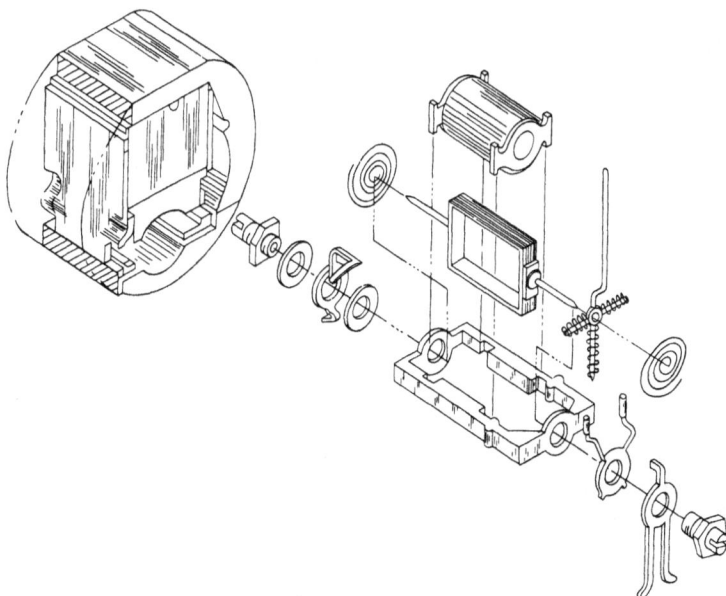

Measuring Direct Current

The moving-coil meter movement can be used to measure any value of direct current by use of *shunt resistors* in parallel with the movement. A diagram of a basic movement with a shunt resistor is shown in Fig. 5-2. The coil of the meter movement will not handle high currents. The typical current rating of a meter movement ranges from 10 µA to 10 mA. This current rating (I_M) is the amount of current needed to move the pointer to the extreme right side of the scale. This is called *full scale*. When this current value is smaller, more turns of wire are used for the coil. More turns allow a strong electromagnetic field to develop. As the current rating (I_M) decreases, the resistance (R_M) of the movement increases because of the greater length of wire. You must

Suspension drawing

Partial view showing air gap

Exploded view of dc pivot and jewel-type meter

FIGURE 5-1 d'Arsonval meter movement.

know the values of I_M and R_M to design a meter circuit that will measure currents higher than the value of I_M.

A shunt resistor is placed in parallel with the meter movement, as shown in Fig. 5-2. Some of the current flowing through the external circuit is "shunted" through the resistor. For this reason the resistor is called a *shunt resistor* (R_{SH}). The value of shunt resistance must be precisely calculated. Shunt resistance produces conditions that allow measurement of ranges of current above the value of the meter movement. The value of shunt resistance is found with the following formula:

$$R_{SH} = \frac{I_M \times R_M}{I_{SH}}$$

where R_{SH} is the shunt resistance in ohms, I_M is the full-scale current of rating of the meter movement in amperes, R_M is the resistance of the meter movement in ohms, and I_{SH} is the current flow through the shunt resistor in amperes.

For example, a 1.0 mA, 100 Ω meter movement can be used to measure 10 mA. The shunt resistance value is found as follows:

$$\begin{aligned} R_{SH} &= \frac{I_M \times R_M}{I_{SH}} \\ &= \frac{0.001 \times 100}{0.009} \\ &= 11.11 \ \Omega \end{aligned}$$

$$1 \ \text{mA} = 0.001 \ \text{A}$$

$$9 \ \text{mA} = 0.009 \ \text{A}$$

The value of I_{SH} is 9 mA (0.009 A), or 10 mA − 1 mA = 9 mA. This amount of current must be shunted through the shunt resistor. At the maximum current of the 10 mA range, I_M equals 1.0 mA. A multirange ammeter is designed with several values of shunt resistance and a switching arrangement, as shown in Fig. 5-3. Multirange meters usually have only one scale. They are read directly or have a multiplying factor so that the scale may easily be read for each range. Scales are discussed in unit 2. This type of scale is called a *linear scale*. It has the same distance between all division marks.

Measuring DC Voltage

Hand-deflection meter movements are another way to measure dc voltage. Meter movements respond to current flow through the electromagnetic coil. As voltage increases, current increases in the same proportion. Voltage values also produce accurate readings on a calibrated meter scale. To measure voltage, a resistor is placed in series with the meter movement. This resistor, shown in Fig. 5-4, is called a multiplier resistor. The purpose of the multiplier is to adjust the value of current across the meter movement. At a certain voltage value ($I_M \times R_M$), full-scale deflection (V_{FS})

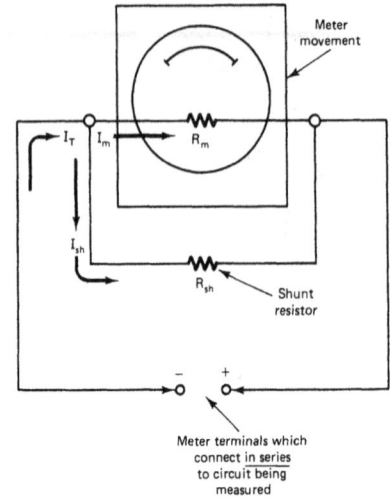

FIGURE 5-2 Meter movement with a shunt resistor to measure current.

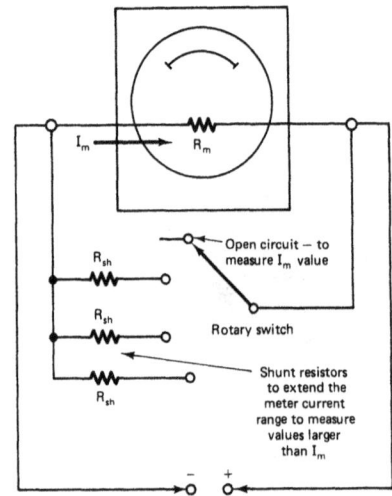

FIGURE 5-3 Switching arrangement for a multirange current meter.

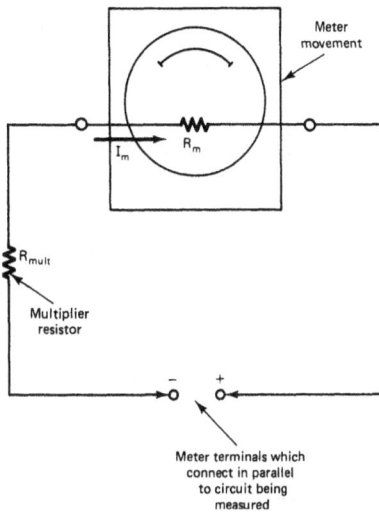

FIGURE 5-4 Meter movement with a multiplier used to measure voltage.

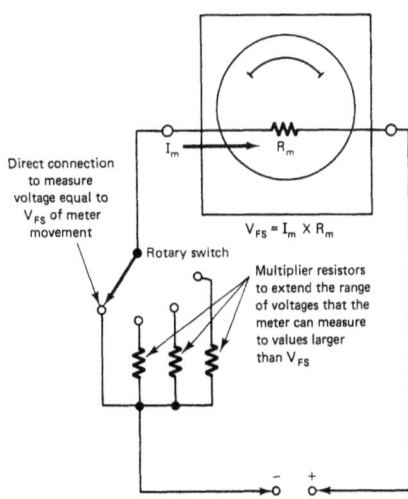

FIGURE 5-5 Switching arrangement for a multirange voltmeter.

on the meter scale occurs. The formula to find full-scale deflection is as follows:

$$V_{FS} = I_M \times R_M$$

A range of dc voltage from 0 to 10 V can be measured with 1.0 mA, 100 Ω meter movement. With 1.0 mA flowing through the 100 Ω meter movement (full scale), 0.1 V is applied across the meter movement ($V_{FS} = 1$ mA $\times$ 100 Ω = 0.1 V). To measure 10 V at full scale, a multiplier resistance is added in series with the meter movement to drop 9.9 V. This value is found with the following formula:

$$R_{mult} = \frac{V_{mult}}{I_T}$$
$$= \frac{9.99}{0.001}$$
$$= 9900 \ \Omega$$

Another way to find the value of a multiplier resistance is to calculate the total resistance in the circuit. The values for the circuit with 10 V applied are found as follows:

$$R_T = \frac{V_T}{I_T}$$
$$= \frac{10\,V}{0.001 \ A}$$
$$= 10,000 \ \Omega$$

The total resistance of the circuit is $R_T = R_M + R_{mult}$ so

$$R_{mult} = R_T - R_M$$
$$= 10,000 \ \Omega - 100 \ \Omega$$
$$= 9900 \ \Omega$$

A switching arrangement is used for several voltage ranges that use the same meter movement. A switching arrangement for a multirange voltmeter is shown in Fig. 5-5.

An important characteristic of meters is *sensitivity*. The sensitivity of a meter increases as the current rating (I_M) decreases. Voltmeters are connected in parallel with a circuit to measure voltage. Part of the circuit current flows through the meter to make the needle deflect. The current that flows through the meter should be very small. More sensitive meters draw less current from the circuit. Sensitivity is measured in ohms per volt and is equal to $1/I_M$. Remember that I_M is the current required for full-scale deflection of the meter. A 1 mA movement would have an ohms per volt rating of 1/0.001, or 1000 Ω/V. This means that the meter would have 1000 Ω of resistance for its 1 V range. Sensitivities of meter movements range from as low as 100 Ω/V to as high as 200,000 Ω/V. Low-sensitivity meters should not be used for making accurate measurements.

Measuring Resistance

Resistance can be measured with a meter movement. An ohm-meter circuit is shown in Fig. 5-6a. The resistance to be measured is connected in series with the meter circuit. A 1.5 V cell is used to supply current to the meter circuit. The scale of the meter is calibrated so that the movement of the pointer indicates a value of resistance. When the meter probes are touched together, the meter pointer is adjusted to full scale. An ohms scale is shown in Fig. 5-7a. The full-scale mark (right side) indicates zero resistance. As higher values of resistance are measured, less current flows through the meter circuit. The ohmmeter pointer deflects less for higher resistance values. The ohms-adjust control is a potentiometer used to adjust the meter pointer. An ohmmeter must be zeroed before measurements are made. The pointer is adjusted to the zero on the scale when the probes are touched together. The purpose of the ohms-adjust control is to allow accurate measurements as the voltage of the battery in the circuit changes. The left side of the scale indicates infinite resistance. This is an open circuit that has no current flow.

In the ohmmeter circuit of Fig. 5-6a a 1.0 mA, 100 Ω meter movement is used. A value of 900 Ω for the current-limiting resistor (R_{lim}) is calculated in Fig. 5-7b. The total resistance of the meter circuit will be $R_M + R_{lim} + R_{ohms}$, or 100 Ω + 900 Ω + 500 Ω = 1500 Ω. When the meter probes are touched together, the total current in the circuit will be 1 mA:

$$I_T = \frac{V_T}{R_T}$$
$$= \frac{1.5 \text{ V}}{1500 \text{ }\Omega}$$
$$= 0.001 \text{ A}$$
$$(1.0 \text{ mA})$$

One milliampere is the current (I_M) for full-scale deflection of the meter. The ohms scale is calibrated to read 0 Ω at the full-scale deflection point of the meter.

The measured resistance that causes half-scale deflection of the meter pointer is equal to the total resistance of the ohmmeter circuit. To measure a 1500 Ω resistance, the resistor is connected between the meter probes. The total circuit resistance is then 3000 Ω. The meter pointer deflects to only one-half of full scale. With twice as much resistance in the circuit, there is only half as much current. The center of the ohms scale is marked 1500 Ω, as shown in Fig. 5-7b.

If the 3000 Ω resistance is measured, the total circuit resistance is 4500 Ω. This is three times the resistance of the meter circuit. The current in the circuit is then only one-third of the

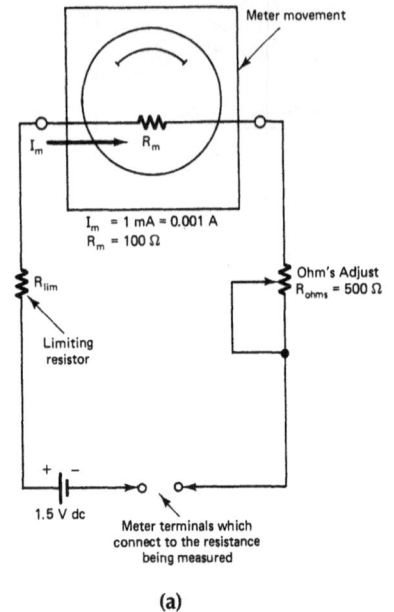

1. Find total resistance (R_T) which is needed in the circuit to cause full-scale deflection of the meter when the meter terminals are touched together:

$$R_T = \frac{V_T}{I_m} = \frac{1.5 \text{ V}}{1 \text{ mA}} = \frac{1.5 \text{ V}}{0.001 \text{ A}} = 1500 \text{ }\Omega$$

2. Find the value of limiting resistor (R_{lim}) needed in the circuit:

$$R_{lim} = R_T - R_{Ohm's \ Adjust} - R_m$$
$$= 1500 \text{ }\Omega - 500 \text{ }\Omega - 100 \text{ }\Omega$$
$$= 900 \text{ }\Omega$$

(b)

FIGURE 5-6 (a) Ohmmeter circuit with a meter movement used to measure resistance. (b) Ohmmeter circuit calculations.

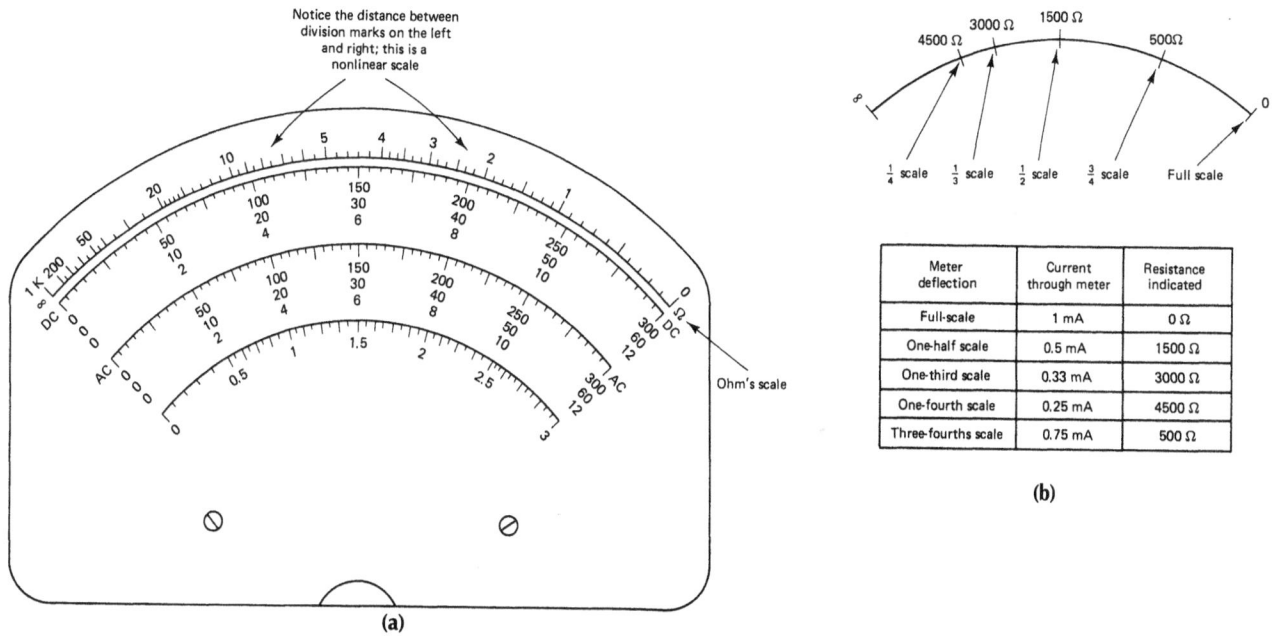

FIGURE 5-7 (a) Ohms scale of a meter. (b) Marking or calibrating an Ohms scale.

Meter deflection	Current through meter	Resistance indicated
Full-scale	1 mA	0 Ω
One-half scale	0.5 mA	1500 Ω
One-third scale	0.33 mA	3000 Ω
One-fourth scale	0.25 mA	4500 Ω
Three-fourths scale	0.75 mA	500 Ω

(b)

full-scale current (.33 mA). The meter pointer deflects to only one-third of full scale when 3000 Ω is measured. The point on the ohms scale for 3000 Ω is marked. The amount of current can be calculated for any value of measured resistance. The ohms scale is marked according to the current value. The table in Fig. 5-7b shows some values used to calibrate the ohms scale of a meter. This type of scale is called a *nonlinear scale*. It has division marks that are farther apart on the right than on the left.

There are limits to the amount of resistance an ohmmeter can measure. Several values can be measured with a multirange ohmmeter that has a small battery (usually 1.5 V) for the ×1 and ×100 ranges and a larger battery (usually 9 or 30 V) for ranges of ×1000 and higher.

Multifunction meters (multimeters) usually are designed to measure voltage, current, and resistance. Most hand-deflection analog meters have one meter movement with internal circuits to make many types of measurements. A rotary switch often is used for multimeters to select the quantity to be measured and the proper range.

Self-Examination

1. The five basic types of instruments are
   ——————, ——————, ——————, ——————,
   ——————.

2. A ———————— meter measures one quantity.

3. The range of a meter movement is extended to measure higher current values by means of addition of a ————————.

4. The following abbreviations are: (a) I_M ―――――,
(b) R_M ―――――, (c) R_{SH} ―――――, and
(d) I_{SH} ―――――.

5. A meter movement is used to measure higher voltages by adding a ―――――.

6. Explain the following abbreviations. (a) V_{FS} ―――――, (b) R_{mult} ―――――.

Solve the following ammeter circuit-design problems.

7. Using a meter movement with resistance (R_M) of 300 Ω and current (I_M) of 400 μA, calculate the value of shunt resistance (R_{SH}) needed to extend its range to measure 200 mA of current. R_{SH} = ―――――

8. If a 200 Ω, 1 mA meter movement is used, what values of shunt resistance are needed to extend its range to measure the following currents?

   a. 5 mA ―――――   b. 10 mA ―――――
   c. 50 mA ―――――   d. 100 mA ―――――
   e. 9.5 mA ―――――   f. 1 A ―――――
   g. 2A ―――――   h. 400 μA ―――――

Solve the following voltmeter circuit-design problems.

9. Find the voltage for full-scale deflection (V_{FS}) of a meter movement with R_M = 400 Ω and I_M = 0.5 mA. V_{FS} = ―――――

10. Find the voltage (V_{mult}) that must be dropped across the multiplier resistor to measure 10 V for the meter movement used in problem 9. V_{mult} = ―――――

11. Find the value of multiplier resistor (R_{mult}) needed in series with the meter movement of problem 9 to measure 10 V. R_{mult} = ―――――

12. Calculate the values of multiplier resistance needed to extend the range of a 1 mA, 100 Ω meter movement to measure the following voltages.

   a. 1 V, R_{mult} = ――――― Ω
   b. 5 V, R_{mult} = ――――― Ω
   c. 15 V, R_{mult} = ――――― Ω
   d. 150 V, R_{mult} = ――――― Ω
   e. 300 V, R_{mult} = ――――― Ω
   f. 500 V, R_{mult} = ――――― Ω
   g. 1000 V, R_{mult} = ――――― Ω

Solve the following ohmmeter circuit-design problems.

13. Use a meter movement with R_m = 500 Ω, I_m = 0.5 mA, and a 1.5 V battery. Find the total resistance needed in the ohmmeter circuit when using a 1.5 V battery. R_T = ―――――

14. Find the value of a limiting resistor (R_{lim}) needed for the circuit in problem 13. R_{lim} = _____

15. Draw a half-circle on a sheet of paper to represent an ohmmeter scale and mark the following points on the scale. Use the circuit values of problems 13 and 14.
    a. Full scale on the right side (0 Ω)
    b. Infinite on the left side (∞Ω)
    c. Half-scale ($R_{lim} + R_M$ value) _____ Ω
    d. One-fourth scale = _____ Ω
    e. Three-fourths scale = _____ Ω

16. Find the value of R_{lim} needed for a 1 mA, 100 Ω meter movement using a 9 V battery. R_{lim} = _____

17. Using the circuit values of problem 16, calculate the values of resistance for the following deflections on an ohmmeter scale:
    a. One-fourth scale = _____ Ω
    b. One-half scale = _____ Ω
    c. Three-fourths scale = _____ Ω

Answers

1. Analog, comparison, CRT, numerical readout, chart recording	2. Single-function
3. Shunt resistance (R_{SH})	4. (a) Meter current; (b) meter resistance; (c) shunt resistance; (d) shunt current
5. Multiplier resistance	6. (a) Full-scale voltage; (b) multiplier resistance
7. R_{SH} = 0.6012 V	8. a. 50 Ω b. 22.22 Ω c. 4.082 Ω d. 2.0202 Ω e. 23.53 Ω f. 0.2002 Ω g. 0.10005 Ω h. No R_{SH}
9. V_{FS} = 0.2 V	10. V_{mult} = 9.8 V
11. R_{mult} = 19,600 Ω	12. a. 900 Ω b. 4900 Ω c. 14,900 Ω d. 149,000 Ω e. 299,900 Ω f. 499,000 Ω g. 999,900 Ω

Answers (continued)

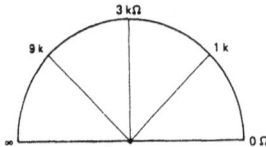

Measuring Electric Power

Electric power is measured with a wattmeter. A meter movement called a *dynamometer* is used for most wattmeters. Figure 5-8 shows a meter movement that has two electromagnetic coils. One coil, called the *current coil*, is connected in series with the circuit to be measured. The other coil, called the *potential coil*, is connected in parallel with the circuit. The strength of each electromagnetic field affects the movement of the meter pointer. The operating principle of this movement is like that of the moving-coil type. The main difference is that an electromagnetic field rather than a permanent magnet field surrounds the moving coil. DC power is found by means of multiplying voltage and current ($P = V \times I$). AC power is found by means of multiplying voltage, current, and the power factor (pf) of the load ($P = V \times I \times \text{pf}$). The true power of an ac circuit is read with a wattmeter. When a load is either inductive or capacitive, the true power is less than apparent power ($V \times I$). The true power is the actual power converted by the load.

Measuring Electric Energy

Wattmeters with dynamometer movements are used to measure electric power. They monitor the voltage and current in a circuit. They are used to measure the electric power converted by a load. The amount of electric energy converted into mechanical energy by a motor load is measured by means of connecting a wattmeter to the motor circuit.

The amount of electric energy used over a period of time is measured with a watt-hour meter. A watt-hour meter normally has a small motor inside. The speed of the motor increases as more current passes through it. The rotor is an aluminum disk and is connected to a numerical display. The display indicates the number of kilowatt-hours of electric energy used. Other types have a numerical display of actual kilowatt-hours used.

Watt-hour meters are connected between the power lines coming into a building and the branch circuits inside the building. The electric energy used by a building must pass through the kilowatt-hour meter. The operation of a watt-hour meter is similar to

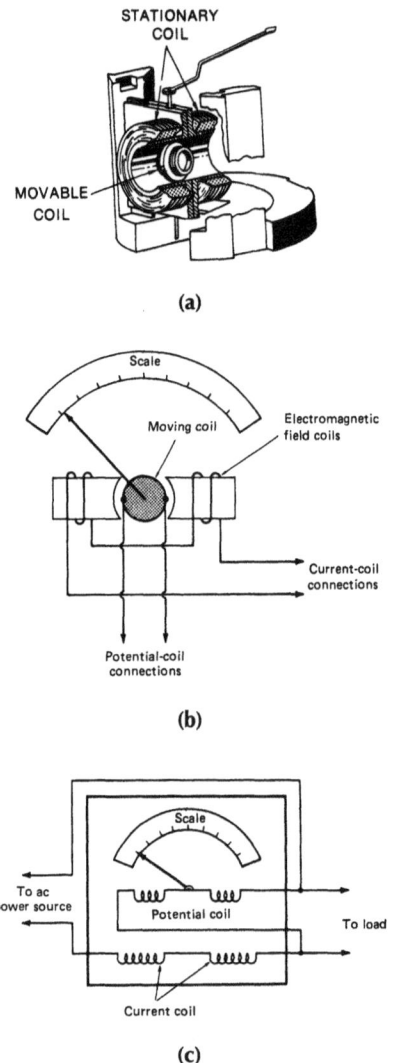

FIGURE 5-8 Meter used to measure electric power. (a) Dynamometer movement. (b) Meter movement schematic. (c) Circuit.

that of a wattmeter. A voltage coil is connected in parallel with power lines to monitor voltage. A current is placed in series with one line to measure current. The voltage and current of the system affect the speed of the aluminum disk. A watt-hour meter is similar to an ac induction motor. This stator is an electromagnet that has two sets of windings: the voltage windings and the current windings. The field developed in the voltage windings causes current to be induced into the aluminum disk. The speed of the disk increases when the voltage or current of the system increases. Because voltage usually stays the same, the speed of the disk increases as current increases. This causes the number display to increase. Watt-hour meters are used to monitor the true power converted in a system over a period of time.

Measuring Three-phase Electric Power

It often is necessary to monitor the three-phase power used by industrial and commercial buildings. A combination of single-phase wattmeters can be used to measure three-phase power, as shown in Fig. 5-9. The method shown is not practical. The sum of the meter readings is used to determine the total power of a three-phase power system. A meter called a *three-phase wattmeter* is used to measure the true power of a three-phase system. The power indicated on the meter depends on the voltage and current of all three phases of the system. Three-phase watt-hour meters are available. They contain three aluminum disks to monitor the three-phase power used by a system over a period of time. The rotation of the shaft is caused by a combination of the power of each phase. The kilowatt-hour display increases as the three-phase power converted by the system increases.

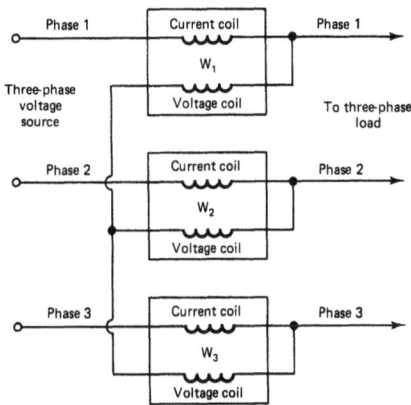

FIGURE 5-9 Use of single-phase wattmeters to measure three-phase power.

Measuring Power Factor

Power factor is the ratio of the measured true power of a system to the apparent power (volts × amperes). A wattmeter, a voltmeter, and an ammeter may be used to find the power factor of a system. The power factor can be found as pf = W/VA. It is more convenient to use a power factor meter when power factor must be measured.

The circuit of a power meter is shown in Fig. 5-10. A power factor meter is similar to a wattmeter. It has two armature coils that rotate because of their electromagnetic field strengths. The armature coils are mounted on the same shaft. They are placed about 90° apart. One coil is connected across the power line in series with a resistance (R). The other coil is connected across the line in series with an inductance (L). The resistor in series with the coil produces a magnetic field from the in-phase part of the power. The inductor in series with the other coil produces a magnetic field caused by the out-of-phase part of the power. The scale of the meter is calibrated to measure power factors of 0 to 1.0, or 0 to 100%.

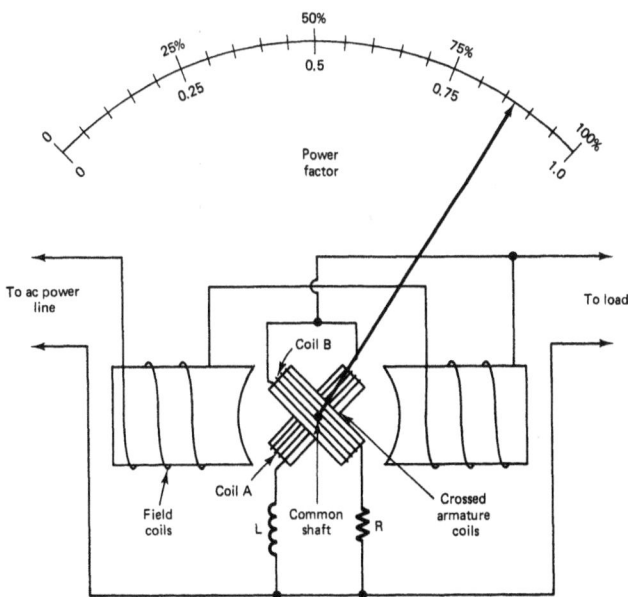

FIGURE 5-10 Circuit of a power factor meter.

Measuring Power Demand

A power-demand meter is an important instrument for industries and large commercial buildings. Power demand is found with the following formula:

$$\text{Power demand} = \frac{\text{peak power used (kW)}}{\text{average power used (kW)}}$$

Power demand is important because it shows the ratio of average power used to the peak value that a utility company must supply. Power demand usually is calculated over 15-, 30-, or 60-min intervals. It may be converted into longer periods of time.

The utility company may penalize industries and businesses if their peak power demand is far greater than their average demand. This is called a *demand charge*. Power-demand meters help companies use their electric power more effectively. High peak demand means the equipment used for power distribution must be larger. The closer the peak demand is to the average demand, the more efficient is the power system.

Measuring Frequency

Another important measurement is *frequency*. The frequency of the power system must remain the same at all times. Frequency refers to the number of cycles of voltage or current that occur in a given period of time. The international unit used to measure frequency is the hertz (Hz). One hertz equals one cycle per second. A table of frequencies is shown in Table 5-1. The standard power frequency in the United States is 60 Hz. Some countries use 50 Hz.

TABLE 5-1 Frequency Values*

Band	Frequency Range
Extremely low frequency	30 Hz–300 Hz
Voice frequency	300 Hz–3 kHz
Very low frequency	3 kHz–30 kHz
Low frequency	30 kHz–300 kHz
Medium frequency	300 kHz–3 MHz
High frequency	3 MHz–30 MHz
Very high frequency	30 MHz–300 MHz
Ultrahigh frequency	300 MHz–3 GHz
Super-high frequency	3 GHz–30 GHz
Extremely high frequency	30 GHz–300 GHz

*Power frequency = 60 Hz; AM radio band = 550–1600 kHz; FM radio band = 88–108 MHz.

Frequency is measured with several different types of meters. An electronic counter is used to measure frequency. Vibrating-reed frequency meters often are used. An oscilloscope also is used to measure frequency. Graphic recording instruments can be used to give a visual display of frequency over a period of time. The power industry, for example, must monitor the frequency of its alternators at all times.

Ground-Fault Indicators

A ground-fault indicator is used to locate faulty grounding of equipment or power systems. Equipment must be properly grounded. A ground-fault indicator is used to check for faulty grounding. Several grounding conditions exist that might be dangerous. Faulty conditions include (1) hot and neutral wires reversed, (2) open equipment-ground wires, (3) open neutral wires, (4) open hot wires, (5) hot and equipment grounds reversed, and (6) hot wires on neutral terminals. Each of these conditions presents a serious problem. Proper wiring eliminates most of these problems. A check with a ground-fault indicator assures safe and efficient electric wiring in a building.

Measuring High Resistance

A megohmmeter is used to measure very high resistances. These resistances are beyond the range of most ohmmeters. Megohmmeters are used to check the quality of insulation on electric equipment in industry. The quality of insulation of equipment varies with age, moisture content, and applied voltage. Megohmmeters are similar to most ohmmeters. They have hand-crank, permanent-magnet dc generators rather than batteries as voltage sources. The operator cranks the dc generator while making a test. Insulation tests should be performed on all power equipment. Insulation breakdown causes equipment to fail. Insulation resistance value can be used to determine when equipment has to be replaced or repaired. A decrease in insulation resistance over time means that a problem might soon exist.

Clamp-on Meters

Clamp-on meters are used to measure current in power lines. They are used to check current by means of clamping around a power line. They are easy to use for maintenance and testing of equipment. The meter is simply clamped around a conductor. Current flow through a conductor produces a magnetic field around the conductor. The magnetic field induces a current into the iron core of the clamp-on part of the meter. The meter scale is calibrated to indicate current flow. An increase in current flow through a power line causes the current induced into the clamp-on part of the meter to increase. Clamp-on meters usually have voltage and resistance ranges to make them more versatile.

Comparison Instruments

Another type of indicator is called a *comparison instrument*. A comparison instrument is used to compare a known value with an unknown value. The accuracy of a comparison instrument is much greater than that of hand-deflection instruments. Comparison instruments can be used to make very accurate measurements.

Wheatstone Bridge

A *Wheatstone bridge* is a comparison instrument. The circuit of a Wheatstone bridge is shown in Fig. 5-11. A voltage source is used with a sensitive zero-centered movement and a circuit called a *resistance bridge.* The resistance circuit has an unknown external resistance (R_x), which is the resistance to be measured. Resistor R_s is the known "standard" resistance. R_s is adjusted so that the current path of R_x and R_s is the same value as the path formed by R_1 and R_2. No current flows through the meter at this time. The meter indicates zero current. This is called a *null condition*. The bridge is said to be balanced. The value of R_s is marked on the instrument to compare with the value of R_x. Resistors R_1 and R_2 are called the *ratio arm* of the bridge circuit. The value of the unknown resistance (R_x) is found with the following formula:

$$R_x = \frac{R_1}{R_2} \times R_s$$

Wheatstone bridges are used to measure resistance with precision. Other comparison instruments are based on the Wheatstone bridge principle. Comparison instruments are used to compare unknown quantities with known quantities in the circuits of the instruments.

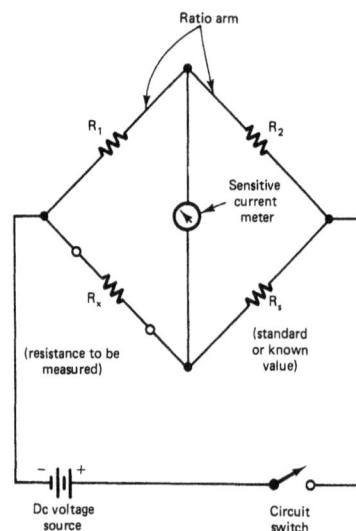

FIGURE 5-11 Circuit of a Wheatstone bridge.

CRT Instruments

CRT instruments usually are called *oscilloscopes*. Oscilloscopes are used to monitor voltages of a circuit visually. The basic part of the oscilloscope is a cathode-ray tube (CRT).

Figure 5-12 shows the construction of a CRT and its internal electron gun assembly. A beam of electrons is produced by the cathode of the tube. Electrons are given off when the filament is heated by a filament voltage. The electrons have a negative (–) charge. They are attracted to the positive (+) potential of anode 1. The number of electrons that pass to anode 1 is changed by the amount of negative (–) voltage applied to the control grid. Anode 2 has a higher positive voltage applied. The higher voltage accelerates the electron beam toward the screen of the CRT. The difference in voltage between anodes 1 and 2 sets the point where the beam strikes the CRT

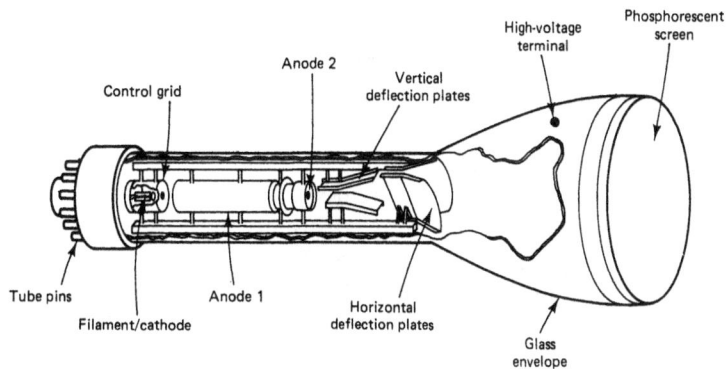

(a)

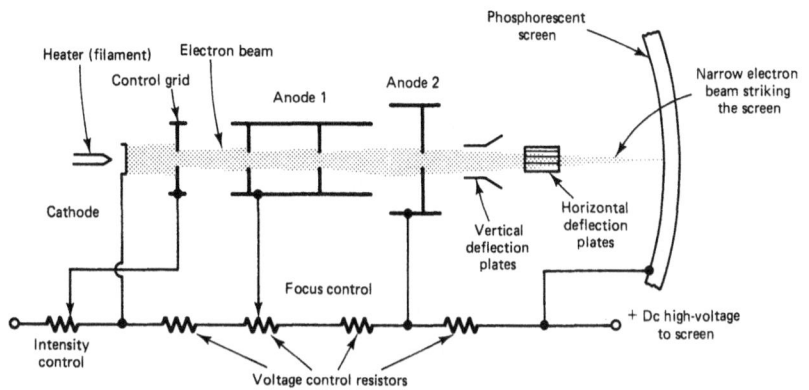

(b)

FIGURE 5-12 Construction of a CRT. (a) Cutaway
view. (b) Electron beam detail.

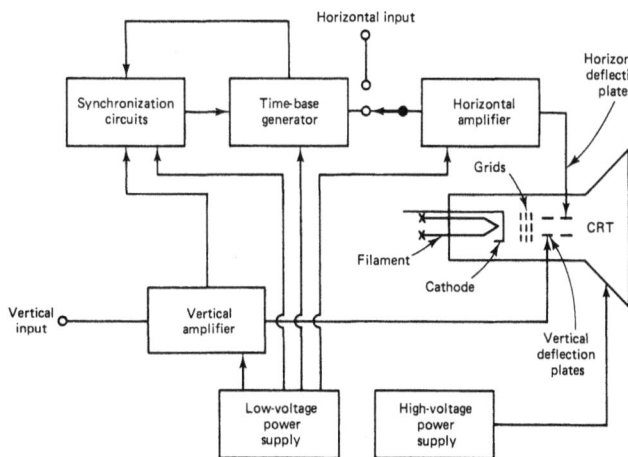

TRFIGURE 5-13 Block diagram of an oscilloscope.

screen. The screen has a phosphorescent coating, which produces light when electrons strike the CRT. The horizontal and vertical movement of the electron beam is controlled with *deflection plates*. With no voltage applied to either set of plates, the electron beam appears as a dot in the center of the CRT screen. The movement of the electron beam caused by the change of plates is called *electrostatic deflection*. When a potential is placed on the horizontal and vertical deflection plates, the electron beam moves. Horizontal deflection is produced by a circuit inside the oscilloscope called a *sweep oscillator circuit*. A voltage "sweeps" the electron beam back and forth across the CRT screen. The horizontal setting of an oscilloscope is adjusted to match the frequency of the voltage being measured. Vertical deflection is caused by the voltage being measured. The voltage to be measured is applied to the vertical deflection of the oscilloscope. A block diagram of an oscilloscope circuit is shown in Fig. 5-13. Oscilloscopes are used to measure ac and dc voltages, frequency, phase relations, distortion in amplifiers, and various timing and special-purpose applications. They also have important medical uses, such as to monitor heartbeat.

Numerical Readout Instruments

Many instruments have numerical readouts to simplify measurement. They make very accurate measurements. Instruments such as digital counters and digital multimeters are commonly used. Numerical readout instruments have internal circuits that produce a digital display of the quantity being measured. These instruments are easy to read because a scale does not have to be interpreted.

Chart-Recording Instruments

Most instruments are used in applications in which no permanent record of the measured quantity is needed. Sometimes instruments must provide a permanent record of a quantity over a period of time. Some chart recorders have pen-and-ink recorders; others have inkless recorders.

A pen-and-ink recorder has a pen that touches a paper chart. The ink leaves a permanent record of the measured quantity on the chart. The chart is either a roll chart, which revolves on rollers under the pen mechanism, or a circular chart, which rotates under the pen. A chart recorder sometimes has more than one pen to record several different quantities at the same time.

The pen of a chart recorder is connected to the meter movement. The pen is supplied with a source of ink. It is moved by the meter movement in the same way as the pointer of a hand-deflection meter. The charts usually have lines that indicate the amount of pen movement. Spaces on the chart are marked according to time periods. Charts are moved under the pen at a constant speed. Spring-drive mechanisms, synchronous ac motors, or dc servomotors are used to drive charts.

Inkless recorders may have voltage applied to the pen point. Heat is produced, which causes a mark to be made on a sensitive paper chart. The advantage of inkless recorders is that ink is not needed. The ink in pen-and-ink recorders must be replaced, and it is often messy to use.

A special type of chart-recording instrument is the X-Y recorder. X-Y recorders are used to plot the relation of two quantities. A variable quantity often is plotted in relation to time. One variable quantity can be plotted in relation to another variable quantity. One input is applied to cause vertical (Y axis) movement of the indicating pen. The other input causes horizontal (X axis) movement of the pen. The plot produced on the chart shows the relation of two quantities connected to the X and Y inputs.

18. Electric power is measured with a _____.

19. Wattmeters with _____ movements are used to measure electric power.

20. The amount of electric energy used over a period of time is measured with a _____ meter.

21. The ratio of true power to apparent power is called _____.

22. The ratio of peak power to average power is _____.

23. A vibrating-reed meter may be used to measure _____.

24. Very high resistances may be measured with a _____.

Solve the following Wheatstone bridge problems.

25. Calculate the values of R_x using $R_1 = 1\ k\Omega$ and $R_2 = 10\ k\Omega$ for the following values of R_s:
    a. $R_s = 100\ \Omega$, $R_x =$ _____ Ω
    b. $R_s = 500\ \Omega$, $R_x =$ _____ Ω
    c. $R_s = 10.2\ k\Omega$, $R_x =$ _____ Ω
    d. $R_s = 5.6\ k\Omega$, $R_x =$ _____ Ω
    e. $R_s = 100\ k\Omega$, $R_x =$ _____ Ω
    f. $R_s = 22\ k\Omega$, $R_x =$ _____ Ω

Answers

18. Wattmeter	19. Dynamometer
20. Watt-hour	21. Power factor
22. Power demand	23. Frequency
24. Megohmmeter	25. a. 10 Ω
	b. 50 Ω
	c. 1020 Ω
	d. 560 Ω
	e. 10,000 Ω
	f. 22,000 Ω

Unit 5 Examination

Electronic Instruments

Instructions: For each of the following, circle the answer that most correctly completes the statement.

1. The type of electric power that is measured in watts is
   a. Apparent power   b. True power
   c. Unity power   d. Power factor

2. The most often used analog meter movement is the
   a. Inclined vane type   b. d'Arsonval Type
   c. Repulsion vane type   d. Solenoid type

3. A device that allows current to flow in only one direction is a
   a. Potentiometer   b. Carbon resistor
   c. Diode   d. Capacitor

4. A meter that measures voltage, current, and resistance is
   a. A single function meter
   b. A multirange meter
   c. A current meter
   d. A multimeter meter

5. The d'Arsonval meter movement rotates as a result of
   a. Mechanical force exerted upon it
   b. Heat energy buildup within the meter
   c. Production of light by cells within the meter
   d. Electric current flowing through it

6. A typical value of current that could pass through a meter movement is
   a. 500 A   b. 1 mA
   c. 10 A   d. 0.0001 µA

7. When a wattmeter is connected into a circuit, it is placed
   a. In parallel with the circuit
   b. In series with the circuit
   c. So that one section of the meter is in series and the other is in parallel with the circuit
   d. So that an excess amount of current will not flow through the circuit under measurement

8. The range of the meter movement of a voltmeter may be extended with
   a. A multiplier resistor in parallel with the movement
   b. A shunt resistor
   c. A multiplier in series with the movement
   d. A parallel potentiometer

9. A meter movement can be used to measure larger values of current by means of
   a. Adding resistors in series with the movement
   b. Adding resistors in parallel with the movement
   c. Using a multiplier
   d. Using overload relays

10. The calculation of a shunt for a meter movement is a good example of the application of
    a. Kirchhoff's current law
    b. Thevinin's theorem
    c. Norton's theorem
    d. The superposition theorem

11. To calculate the value of R_{SH} in a meter shunt circuit, which equation should be used?

    a. $R_{SH} = \dfrac{I_{SH} \times R_M}{I_M}$

    b. $R_{SH} = \dfrac{I_{SH} \times V_{SH}}{R_M}$

    c. $R_{SH} = \dfrac{I_M \times R_M}{I_{SH}}$

    d. $R_{SH} = \dfrac{I_{SH} \times I_M}{R_M}$

12. The "loading" effect of a meter is less evident when a meter is used where the ohms-per-volt rating is
    a. 20,000 Ω/V b. 1000 Ω/V
    c. 0 Ω/V d. 10,000 Ω/V

True-False: Place either T or F in each blank.

_____ 13. Electric power is measured with a dynamometer meter movement.

_____ 14. Watt-hour meters are used to monitor electric energy used in homes.

_____ 15. The measurement of power factor provides a ratio of true power and apparent power in a system.

_____ 16. Power demand is measured as a ratio of peak power and potential power used by a system.

_____ 17. Frequency may be measured with a VOM.

_____ 18. Clamp-on meters are versatile instruments that may be used to measure current in power lines.

_____ 19. Low values of resistance are typically measured with a megohmmeter.

_____ 20. A Wheatstone bridge is a type of comparison instrument.

Inductance and Capacitance

Two very important electronic properties are inductance and capacitance. This unit discusses these properties and inductors and capacitors. Time-constant circuits, which rely on inductance and capacitance also are discussed.

UNIT OBJECTIVES

1. Explain how a capacitor operates.

2. Solve *RC* time-constant problems.

3. Define inductance and inductors.

4. Define capacitance and capacitors.

5. List the factors determining capacitance.

6. Describe the construction of various types of capacitors.

7. Solve *RC* time-constant problems.

8. Calculate total capacitance of capacitors in various series, parallel, and combination configurations.

9. Calculate total inductance of inductors in various series, parallel, and combination configurations.

10. Explain what an inductor is and how it operates.

11. List factors that affect inductance.

12. Identify different types of inductors.

13. Explain the concept of mutual inductance.

Important Terms

Review the following terms to gain an understanding of some of the topics discussed in this unit.

Air-core inductor A coil wound on an insulated core or a coil of wire that does not have a metal core.

Capacitance (C) The property of a device to oppose changes in voltage caused by energy stored in its electrostatic field.

Capacitor A device that has capacitance and usually is made of two metal plate materials separated by a dielectric material (insulator).

Choke coil An inductor coil used for power-supply applications.

Condenser A term occasionally used to mean *capacitor*.

Decay A term used to describe a gradual reduction in value of a voltage or current.

Decay time The time for a capacitor to discharge to a certain percentage of its original charge or the time required for current through an inductor to reduce to a percentage of its maximum value.

Dielectric An insulating material placed between the metal plates of a capacitor.

Dielectric constant A number that represents the ability of a dielectric to develop an electrostatic field; compared with air, which has a value of 1.0.

Electrolytic capacitor A capacitor that has a positive plate made of aluminum and a negative plate of dry paste or liquid.

Electrostatic field The field developed around a material because of the energy of an electrical charge.

Farad The unit of measurement of capacitance required to contain a charge of 1 C when a potential of 1 V is applied.

Henry (H) The unit of measurement of the inductance produced when a voltage of 1 V is induced when the current through a coil is changing at a rate of 1 A/s.

Inductance (L) The property of a circuit to oppose changes in current caused by energy stored in a magnetic field.

Inductive circuit A circuit that has one or more inductors or has the property of inductance, such as an electric motor circuit.

Inductor A coil of wire that has the property of inductance and is used in a circuit for that purpose.

Mutual inductance (M) A state that occurs when two coils are located close together, so that the magnetic flux of each coil affects the other in terms of its inductance properties.

Time constant (*RC*) The time required for the voltage across a capacitor in an *RC* circuit to increase to 63% of its maximum value or decrease to 37% of its maximum value; time = *L/R*.

Working voltage A rating of capacitors that is the maximum voltage that can be placed across the plates of a capacitor without damage.

Inductance

When energized with dc voltage, a coil produces a magnetic field around itself. DC current flow produces a constant magnetic field around a coil. When the same coil is supplied with ac voltage, the constantly changing ac current produces a constantly changing magnetic flux. This changing flux sets up a magnetic field with a constantly reversing polarity and changing strength. It also induces a counterelectromotive force (CEMF) or counter voltage. This counterelectromotive force opposes the source voltage. CEMF limits current flow from the source.

The opposition to the flow of ac current by a magnetic field is called *inductance* (*L*). The opposition to current flow of an inductive device depends on the resistance of the wire and the magnetic properties of the circuit. The opposition caused by the magnetic effect in ac circuits is called *inductive reactance* (X_L). X_L varies with the applied frequency and is found with the formula $X_L = 2\pi \times f \times L$, where $2\pi = 6.28$, *f* is the applied frequency in hertz, and *L* is the inductance in henrys. The basic unit of inductance is the henry (H). Inductance in ac circuits is discussed in a companion text—*Understanding AC Circuits*.

At zero frequency (or dc), there is no opposition caused by inductance. Only the resistance of a coil limits current flow. As ac frequency increases, the inductive effect becomes greater. Many ac machines have magnetic circuits in one form or another. The inductive reactance of an ac circuit usually has more effect on current flow than resistance. An ohmmeter measures dc resistance only. Inductive reactance must be calculated or determined experimentally.

Figure 6-1 shows symbols used to indicate various types of inductors. Inductors are all coils of wire designed for specific functions. Often they are called *choke coils*. Choke coils are used to pass dc current and block ac current flow.

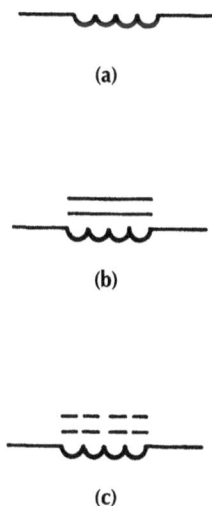

(a)

(b)

(c)

FIGURE 6-1 Symbols used for inductors. (a) Air core. (b) Iron core. (c) Powdered metal core.

Capacitance

When two conductors are separated by an insulator (dielectric), an electrostatic charge may be set up on the conductors. This charge becomes a source of stored energy. The strength of the charge depends on the applied voltage, the size of the conductors, which are called *plates,* and the quality, or dielectric strength, of the insulation. The closer the two plates are placed together, the more charge may be set up on them. This type of device is called a *capacitor.* The size of capacitors is measured in units of farads, microfarads, and picofarads. Capacitor symbols are shown in Fig. 6-2.

When a dc voltage is applied to the plates of a capacitor, the capacitor charges to the value of the source voltage. The dielectric between the plates then stops current flow. The capacitor remains charged to the value of the dc voltage source. When the voltage source is removed, the capacitor charge leaks away. DC current flows to or from a capacitor only when the source voltage is turned on or off.

If the same capacitor is connected to an ac voltage source, the voltage constantly changes. The capacitor receives energy from the source during one-quarter cycle, and very little current flows through the dielectric of a capacitor. AC voltage applied to a capacitor causes it constantly to change its amount of charge. Capacitors have the ability to pass ac current because of their charging and discharging action. They can be used to pass ac current and block dc current flow. The opposition of a capacitor to a source voltage depends on frequency. The faster the applied voltage changes across a capacitor, the more easily the capacitor passes current. Capacitance in ac circuits is discussed in a companion text—*Understanding AC Circuits.*

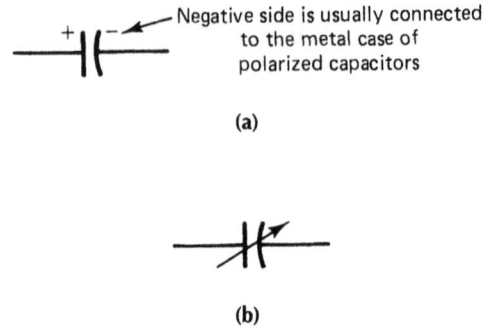

(a)

(b)

FIGURE 6-2 Capacitor symbols. (a) Fixed capacitor. (b) Variable capacitor.

Inductive Effects in Circuits

Inductance is the property of a circuit that opposes changes in current flow. Inductance is the characteristic of an electrical circuit that opposes changes in current flow. Any coil of wire has inductance. Coils oppose current changes by producing a countervoltage (CEMF). The term *reactive* is used because of the reaction of a coil to changes in applied voltage. The opposition of a coil to ac is called *reactance* (X) and is measured in ohms. The subscript L is added to represent inductive reactance (X_L). Inductive reactance depends on the "speed," or frequency, of the ac source.

Inductance does not change with changes in ac frequency. Coils have inductance because of their ability to oppose current change. This property stays the same for any coil. The factors that determine inductance are the number of turns of the coil, the area and length of the core material, and the type of core used. The main factor is the *number of coil turns.* Inductive reactance (X_L) does change with changes in frequency. This can be verified with the formula $X_L = 2\pi fL$. As frequency is increased, inductive reactance increases. Inductors have a low *dc resistance.* This value can be measured with an ohmmeter.

Mutual Inductance

When inductors are connected together, a property called *mutual inductance* (M) must be considered. Mutual inductance is the magnetic field interaction or flux linkage between coils. The amount of flux linkage is called the *coefficient of coupling* (k).

If all the lines of force of one coil cut across a nearby coil, *unity coupling* exists. The many possibilities are determined by coil placement and coupling between coils. The amount of mutual inductance between coils is found with the following formula:

$$\text{Mutual inductance } (M) = k \times L_1 \times L_2$$

The term k is the coefficient of coupling, which gives the amount of coupling. L_1 and L_2 are the inductance values of the coils. Mutual inductance should be considered when two or more coils are connected together.

Inductors in Series and Parallel

Inductors may be connected in series or parallel. When inductors are connected to prevent the magnetic field of one from affecting the others, the following formulas are used to find total inductance (L_T):

1. Series inductance:

$$L_T = L_1 + L_2 + L_3 + \ldots + L_n$$

2. Parallel inductance:

$$\frac{1}{L_T} = \frac{1}{L_1} + \frac{1}{L_2} + \frac{1}{L_3} + \ldots + \frac{1}{L_T}$$

L_1, L_2, L_3, and so on, are inductance values in henrys.

When inductors are connected so that the magnetic field of one affects the other, mutual inductance increases or decreases the total inductance. The effect of mutual inductance depends on the physical positioning of the inductors. The distance apart and the direction in which the coils are wound affects mutual inductance. Inductors are connected in series or parallel with an aiding or opposing mutual inductance (M). The formulas used to find total inductance (L_T) are as follows:

1. Series aiding:

$$L_T = L_1 + L_2 + 2M$$

2. Series opposing:

$$L_T = L_1 + L_2 - 2M$$

3. Parallel aiding:

$$\frac{1}{L_T} = \frac{1}{L_1 + M} + \frac{1}{L_2 + M}$$

4. Parallel opposing:

$$\frac{1}{L_T} = \frac{1}{L_1 - M} + \frac{1}{L_2 - M}$$

L_1 and L_2 are the inductance values, and M is the value of mutual inductance.

Inductive Current Relations

A change in current through a coil causes a change in the magnetic flux around the coil. The CEMF of the coil is at maximum value when current value changes. Because CEMF opposes source voltage, it is directly opposite applied voltage.

Lenz's law states that the countervoltage (CEMF) always opposes a change in current. CEMF opposes the rise in current and causes current to decrease. When the values of the countervoltage and the applied voltage are maximum, the circuit current is maximum. The dc resistance of a coil is small. Ohm's law can be used with inductive dc circuits to find current as follows:

$$I_L = \frac{V_L}{R}$$

where I_L is the current through the coil, V_L is the voltage across the coil, and R is the dc resistance of the coil (in ohms).

Capacitive Effects in Circuits

Inductance is defined as the property of a circuit to oppose changes in current. Capacitance is the property of a circuit to oppose changes in voltage. Inductance stores energy in an electromagnetic field, whereas capacitance stores energy in an electrostatic field. A capacitor is measured in a unit called the farad. One farad is the amount of capacitance that allows a current of 1 A to flow when the voltage change across the plates of a capacitor is 1 V/s. The farad is too large for practical use. The microfarad (0.000001 F, abbreviated μF or MFD) is the most common subunit of capacitance. For high-frequency ac circuits, the microfarad also is too large. The unit micromicrofarad (0.000,000,000,001 μF, abbreviated μμF or MMFD) is used. This unit is called a picofarad (pF) to avoid confusion.

The three factors that determine the capacitance of a capacitor are as follows:

1. *Plate area.* Increasing plate area increases capacitance.

2. *Distance between plates.* Capacitance decreases when the distance between the plates increases.

3. *Dielectric material.* Dielectrics, including air, are used as insulators between capacitor plates. They are rated with a dielectric constant (k). Capacitors with higher dielectric constants have higher capacitance.

Capacitors in Series

Adding capacitors in series has the same effect as increasing the distance between plates. This reduces total capacitance. The total capacitance of capacitors connected in series is less than any individual capacitance. Total capacitance (C_T) is found in the same way as parallel resistance. The reciprocal formula is used, as follows:

$$\frac{1}{C_T} = \frac{1}{C_1} + \frac{1}{C_2} + \frac{1}{C_3} + \ldots + \frac{1}{C_N}$$

Capacitors in Parallel

Capacitors in parallel are similar to one capacitor with its plate area increased. Doubling the plate area doubles capacitance. Capacitance in parallel is found by adding individual values, just as with series resistors:

$$C_T = C_1 + C_2 + C_3 + \ldots + C_N$$

Capacitor Charging and Discharging

Refer to the circuit of Fig. 6-3. A capacitor is shown connected to a two-position switch. With the switch in position 1, the capacitor is uncharged. No voltage is applied to the capacitor. Each plate of the capacitor is neutral because no voltage is applied. When a voltage is placed across the capacitor, an electrostatic field or charge is developed between its plates. When the switch is placed in position 2 the capacitor is placed across a 6 V battery. This causes displacement of electrons in the circuit. Electrons accumulate on the negative plate of the capacitor. At the same time, electrons leave the positive plate. This causes a difference of potential to develop across the capacitor. When electrons move onto the negative plate, the negative plate becomes more negative. As electrons leave the positive plate, the positive plate becomes more positive.

The polarity of the potential that exists across the capacitor opposes the source voltage. As the capacitor continues to charge, the voltage across the capacitor increases. Current flow stops when the voltage across the capacitor is equal to the source voltage. The two voltages cancel each other out. No current actually flows through the capacitor because the material between the plates of a capacitor is an insulator.

An important safety factor to remember is that a capacitor may hold a charge for a long time. Capacitors can be an electrical shock hazard if not handled properly.

When a capacitor is discharged, the charges on the plates become neutralized. A current path between the two plates must be developed. When the switch in Fig. 6-3 is moved to position 1, the electrons on the negative plate move to the positive plate and neutralize the charge. The capacitor releases the energy it has absorbed during its charging as it discharges through the resistor.

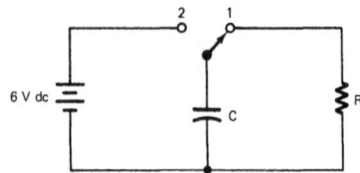

FIGURE 6-3 Capacitor charge and discharge circuit.

Types of Capacitors

Capacitors are classified as either *fixed* or *variable*. Fixed capacitors have one value of capacitance. Variable capacitors are constructed to allow capacitance to vary over a range of values. Variable capacitors often use air as the dielectric. Changing the position of the movable plates varies the capacitance. This changes the plate area of the capacitor. When the movable plates are fully meshed together with the stationary plates, the capacitance is maximum.

Fixed capacitors come in many types. Some types of fixed capacitors are as follows:

1. *Paper capacitors.* Paper capacitors have paper as the dielectric. As shown in Fig. 6-4, they are made of flat strips of metal foil plates separated by a dielectric, which is usually waxed paper. Paper capacitors have values in the picofarad and low-microfarad ranges. The voltage ratings usually are less than 600 V. Paper capacitors usually are sealed with wax to prevent moisture problems.

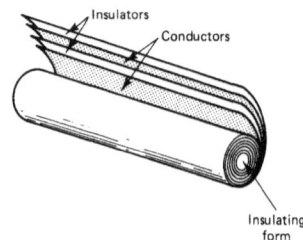

FIGURE 6-4 Paper capacitor construction.

The voltage rating of capacitors is very important. A typical set of values marked on a capacitor might be 10 μF, 50 DCWV. This capacitor has a capacitance of 10 μF and a dc working voltage of 50 V. This means that a voltage in excess of 50 V could damage the plates of the capacitors.

2. *Mica capacitors.* Mica capacitors have a layer of mica and a layer of plate material. The capacitance is usually small (in the picofarad range). They are small in physical size but have high voltage ratings.

3. *Oil-filled capacitors.* Oil-filled capacitors are used when high capacitance and high voltage ratings are needed. They resemble paper capacitors immersed in oil. When soaked in oil the paper has a high dielectric constant.

4. *Ceramic capacitors.* Ceramic capacitors have a ceramic dielectric. The plates are thin films of metal deposited on ceramic material or made in the shape of a disk. They are covered with a moisture-proof coating and have high voltage ratings.

5. *Electrolytic capacitors.* Electrolytic capacitors are used when very high capacitance is needed. Electrolytic capacitors contain a paste electrolyte. They have two metal plates with the electrolyte between them and usually are housed in a cylindrical aluminum can. The aluminum can is the negative terminal of the capacitor. The positive terminal (or terminals) is brought out of the can at the bottom. The size and voltage rating usually are printed on the capacitor. Electrolytic capacitors often have two or more capacitors housed in one unit. They are called *multisection capacitors.*

The positive plate of an electrolytic capacitor is aluminum foil covered with a thin oxide film. The film is formed by an electrochemical reaction and acts as the dielectric. A strip of paper that contains a paste electrolyte is placed next to the positive plate. Another strip of aluminum foil is placed next to the electrolyte. This strip is the negative plate of the capacitor. These three layers are then coiled up and placed into a cylinder.

Electrolytic capacitors are said to be *polarized.* If the positive plate is connected to the negative terminal of a voltage source, the capacitor becomes short-circuited. Capacitor polarity is marked on the capacitor to prevent this from happening. Special high-value nonpolarized electrolytic capacitors are used for ac applications. Motor-starting capacitors are an example.

Time-Constant Circuits

Time-constant circuits have several applications in electronics. They are used in many industrial processes and computer and communication circuit design. Timing functions may be as simple as on-off control or as complex as elaborate sequential operations. The controlled element of a timing circuit can be a motor, a relay, a solenoid, a lamp, or some other circuit or device.

Time-constant circuits use the properties of inductance or capacitance to operate as timing circuits to control load devices. Inductance (L) in circuits opposes changes in current, and capacitance (C) opposes changes in voltage. The reaction time of inductors and capacitors to oppose changes of current or voltage depends on the resistance (R) in the circuit. The time (in seconds) for a capacitor or an inductor to react is called its *time constant.*

RL Time Constants

In Fig. 6-5a a resistor and an inductor are connected in series to a voltage source. Current rises from zero to maximum after a certain time period because of the CEMF of the magnetic field surrounding the inductor. The time required for the current to reach maximum is controlled by the values of R and L. Resistance opposes current flow, and inductance opposes *changes* in current. The change in current occurs from its zero to maximum values. The time required for the current to reach about 63% of maximum is found with the formula $t = L/R$. Time constant (t) is in seconds, L is in henrys, and R is in ohms. After one time constant, current through the inductor is about 63% of maximum (Fig. 6-6a). After five time constants, current through the inductor is maximum. This is an *exponential* relation.

Assume that the maximum current (I_{max}) in the circuit is 100 mA and the time required for the current to reach 63% of 100 mA is 1.0 s. The time required for the current to reach the maximum (100 mA) is 5 s ($t \times 5 = 1.0 \times 5 = 5$ s).

The current flow through the inductor (I_L) varies after the circuit switch is turned on until it reaches its maximum value after five time constants. The values of current are as follows:

After 1 time constant: 63% × 100 mA = 63 mA

After 2 time constants: 86% × 100 mA = 86 mA

After 3 time constants: 95% × 100 mA = 95 mA

After 4 time constants: 98% × 100 mA = 98 mA

After 5 time constants: 100% × 100 mA = 100 mA

After five time constants (5 s for this example), I_L is approximately equal to its maximum value. The buildup of I_L in an *RL* circuit is called its *charging* condition. Charging occurs until I_L stabilizes at its maximum value, as shown in Fig. 6-6a.

RL circuits also have a *discharge* condition. When the switch of Fig. 6-5a is opened, I_L decreases at a rate determined by the circuit time constant. Assume that a wire is placed across the points where the power source was connected at the same time the switch is turned off. The current through the inductor (I_L) discharges, as shown in Fig. 6-6b, until it stabilizes. The approximate values of current during discharge are as follows:

After 1 time constant: 37% × 100 mA = 37 mA

After 2 time constants: 14% × 100 mA = 14 mA

After 3 time constants: 5% × 100 mA = 5 mA

After 4 time constants: 2% × 100 mA = 2 mA

After 5 time constants: 0% × 100 mA = 0 mA

RC Time Constants

A resistor and capacitor are connected in series to a voltage source in Fig. 6-5b. The time for the voltage across the capacitor to reach maximum is controlled by the values of R and C. Resistance opposes current flow in the circuit, whereas capacitance opposes changes in voltage. The change in voltage across the capacitor is from zero to maximum. The amount of time for a capacitor to charge to about 63% of the applied voltage is found by using the formula $t = R \times C$. Time (t) is in seconds, R is in ohms, and C is in farads.

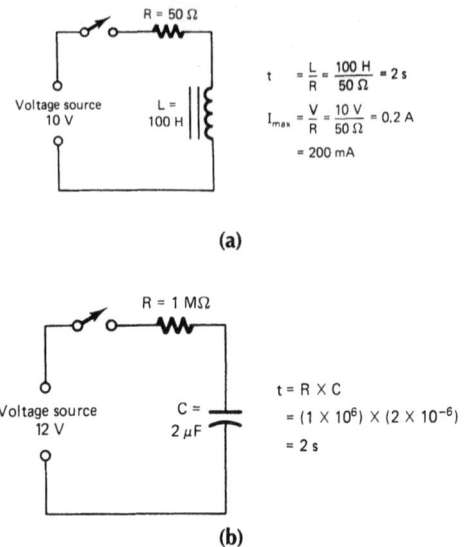

$$t = \frac{L}{R} = \frac{100\ H}{50\ \Omega} = 2\ s$$

$$I_{max} = \frac{V}{R} = \frac{10\ V}{50\ \Omega} = 0.2\ A$$

$$= 200\ mA$$

(a)

$$t = R \times C$$
$$= (1 \times 10^6) \times (2 \times 10^{-6})$$
$$= 2\ s$$

(b)

FIGURE 6-5 Time constant circuits. (a) *RL* circuit. (b) *RC* circuit.

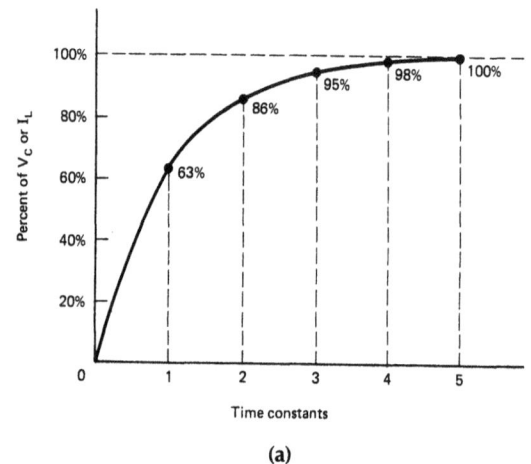

(a)

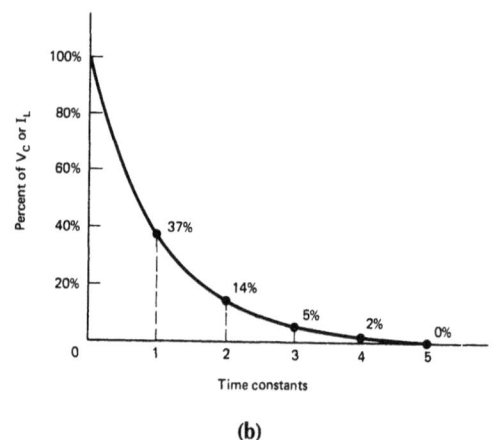

(b)

FIGURE 6-6 Universal time constant curves. (a) Charging or rise time curve. (b) Discharging or decay time curve.

After one time constant, the voltage across the capacitor is about 63% of the source voltage. Figure 6-6a shows a time versus voltage charging curve. In five time constants, the voltage across the capacitor approximately equals the source voltage. The time for the capacitor of Fig. 6-5b to charge to 63% of the source voltage is one time constant, or 3 s. The time for the voltage across the capacitor to equal the source voltage is 15 s ($t \times 5 = 3 \times 5 = 15$ s). This is an *exponential* relation.

The voltage across the capacitor (V_C) in an *RC* circuit increases until it reaches the source voltage value. When the switch in Fig. 6-5b is closed, V_C is approximately 12 V after five time constants. The approximate values of voltage across the capacitor are as follows:

After 1 time constant: 63% × 12 V = 7.56 V

After 2 time constants: 86% × 12 V = 10.32 V

After 3 time constants: 95% × 12 V = 11.40 V

After 4 time constants: 98% × 12 V = 11.76 V

After 5 time constants: 100% × 12 V = 12.00 V

After five time constants (15 s for this example), V_C is approximately equal to the source voltage. The buildup of voltage across the capacitor is called its *charging condition,* as shown in Fig. 6-6a.

Discharging of the capacitor is similar to charging (see Fig. 6-6b). When the source voltage is removed, the capacitor discharges about 63% of the source voltage in one time constant. In five time constants a capacitor discharges 100% of its voltage. When the switch of Fig. 6-5a is opened, V_C decreases at a rate determined by the circuit time constant. Assume that a wire is placed across the points where the power source was connected at the same time the switch is turned off. The voltage across the capacitor (V_C) discharges, as shown in Fig. 6-6b, until it reaches minimum value. The approximate values of capacitor voltage during discharge are as follows:

After 1 time constant: 37% × 12 V = 4.44 V

After 2 time constants: 14% × 12 V = 1.68 V

After 3 time constants: 5% × 12 V = 0.60 V

After 4 time constants: 2% × 12 V = 0.24 V

After 5 time constants: 0% × 12 V = 0.00 V

Universal Time-Constant Curves

The time for current changes in inductive circuits and voltage changes in capacitive circuits is similar. The curves shown in Fig. 6-6 are called *universal time-constant curves*. The vertical axis shows percentages of capacitor voltage (V_C), or percentages of inductor current (I_L). The horizontal axis shows time-constant values. Sometimes the charging curve is called the *rise time* and the discharging curve is called the *decay time*. Each of these is an *exponential* relation.

These curves are useful because they contain the voltage or current values for *RC* and *RL* circuits to be determined at any given time. For example, the charge on a capacitor of a series *RC* circuit after 2.5 time constants is between 8.6 and 9.5 V. A close estimation of voltage across the capacitor may be found by means of interpolation of values, as follows:

$$V_C = 8.6 + \frac{9.5 - 8.6}{2}$$
$$= 8.6 + 0.45$$
$$= 9.05 \ V$$

The charge on a capacitor after a given period of time can be estimated on a universal time-constant curve. For example, after 25 s, what is the approximate charge on the capacitor of Fig. 6-5b? This is a difference of 2.3 V. The value of 25 s is 10 s after the first time constant (25 s = 15 + 10). The time is then $^{10}\!/_{15}$, or cf $^2\!/_3$, of the distance from the first time constant (15 s). The approximation of V_C after 25 s is as follows:

$$V_C = 6.3 + \frac{2}{3} \times 2.3$$
$$= 6.3 + 1.53$$
$$= 7.83 \ V$$

Values along the exponential universal time-constant curve obtained with the methods described are approximations. The exact values may be obtained by means of calculus procedures and natural logarithms. Such approximations are commonly used for electronic circuit design, in which extremely precise values are not necessary.

Self-Examination

Instructions: Solve the following inductance problems by placing the correct answer in the blank.

1. Total inductance in series: $L_1 = 2$ H, $L_2 = 3$ H, $L_3 = 2$ H, $L_T =$ _____

2. Total inductance in parallel: $L_1 = 2$ H, $L_2 = 3$ H, $L_3 = 8$ H, $L_T =$ _____

3. Mutual inductance in series to increase inductance (fields aiding):
   a. $L_1 = 2$, $L_2 = 5$, $M = 0.55$, $L_T =$ _____
   b. $L_1 = 3$, $L_2 = 2$, $M = 0.35$, $L_T =$ _____

4. Mutual inductance in series to decrease inductance (fields opposing):
   a. $L_1 = 4$, $L_2 = 3$, $M = 0.85$, $L_T =$ _____
   b. $L_1 = 3$, $L_2 = 3$, $M = 0.4$, $L_T =$ _____

5. Compute problems 3 and 4 for parallel inductance.

   3a. _____   4a. _____

   3b. _____   4b. _____

Solve each of the following *capacitance* problems.

6. Compute the following for total capacitance in series:

   a. $C_1 - 10\ \mu F$

       $C_2 - 10\ \mu F$

       $C_3 - 30\ \mu F$

       $C_T =$ _____

   b. 40 μF

       20 μF

       20 μF

       $C_T =$ _____

   c. 80 μF

       60 μF

       80 μF

       $C_T =$ _____

7. Compute the following for total capacitance in parallel:

   a. $C_1 = 80\ \mu F$

       $C_2 = 60\ \mu F$

       $C_3 = 80\ \mu F$

       $C_T =$ _____

   b. 50 μF

       20 μF

       30 μF

       $C_T =$ _____

   c. 70 μF

       50 μF

       50 μF

       $C_T =$ _____

Solve the following problems, which deal with *RL* time constant circuits.

8. Refer to Fig. 6-5a. With the following values of R and L, calculate the *RL* time constant (t) of the circuits.

   a. $R = 500\ \Omega$, $L = 200\ mH$, $t =$ _____

   b. $R = 10\ k\Omega$, $L = 15\ H$, $t =$ _____

   c. $R = 2.5\ k\Omega$, $L = 60\ \mu H$, $t =$ _____

   d. $R = 200\ \Omega$, $L = 10\ H$, $t =$ _____

   e. $R = 50\ k\Omega$, $L = 10\ mH$, $t =$ _____

9. In a series *RL* circuit with a maximum current of 200 mA, what are the approximate values of current through the circuit after the following time constants during its charging condition?

a. $t = 1.5$, $I =$ _____

b. $t = 2.0$, $I =$ _____

c. $t = 3.0$, $I =$ _____

d. $t = 3.5$, $I =$ _____

e. $t = 4.8$, $I =$ _____

10. In a series *RL* circuit with a maximum current of 30 mA, what are the approximate values of current through the circuit (I_L) after the following time constants during its discharging condition?

a. 1.0   b. 1.5   c. 2.6

d. 3.2   e. 5.0

Solve the following problems, which deal with *RC* time-constant circuits.

11. Refer to Fig. 6-5b. With the following values *R* and *C*, calculate the *RC* time constant (*t*) of the circuits.

a. $R = 1.5$ MΩ, $C = 1.0$ μF, $t =$ _____

b. $R = 10$ kΩ, $C = 0.002$ μF, $t =$ _____

c. $R = 47$ kΩ, $C = 0.01$ μF, $t =$ _____

d. $R = 50$ kΩ, $C = 10$ pF, $t =$ _____

e. $R = 1$ kΩ, $C = 150$ μF; $t =$ _____

12. In a series *RC* circuit with a maximum voltage of 210 V, what are the approximate values of voltage across the capacitor after the following time constants during its charging conditions?

a. $t = 1.5$, $I =$ _____

b. $t = 2.0$, $I =$ _____

c. $t = 3.0$, $I =$ _____

d. $t = 3.5$, $I =$ _____

e. $t = 4.8$, $I =$ _____

13. In a series *RC* circuit with a maximum voltage of 40 V, what are the approximate values of V_C after the following time constants during its discharging condition?

a. $t = 1.0$, $V_C =$ _____

b. $t = 1.5$, $V_C =$ _____

c. $t = 2.6$, $V_C =$ _____

d. $t = 3.2$, $V_C =$ _____

e. $t = 4.4$, $V_C =$ _____

Answers

1. 7 H

2. 1.043 H

3. a. 8.1 H; b. 5.7 H

4. a. 5.3 H; b. 5.2 H

5. 3a. 1.748 H; b. 1.38 H

   4a. 1.278 H; b. 1.3 H

6. a. 4.29 μF; b. 8 μF;

   c. 24.04 μF

7. a. 220 μF; b. 100 μF;

   c. 170 μF

8. a. 0.0004 s; b. 0.0015 s;

   c. 2.4×10^{-8} s; d. 0.05 s;

   e. 0.2 μs

9. a. 150 mA; b. 172 mA;

   c. 190 mA; d. 193 mA;

   e. 199 mA

10. a. 11.1 mA; b. 7.5 mA;

    c. 2.1 mA; d. 1.2 mA;

    e. 0 mA

11. a. 1.5 s; b. 2×10^{-5} s;

    c. 4.7×10^{-4} s; d. 5×10^{-7} s;

    e. 1.5×10^{-1} s

12. a. 155.4 V; b. 180.6 V;

    c. 199.5 V; d. 204.75 V;

    e. 209.16 V

13. a. 14.8 V; b. 10 V; c. 2.8 V;

    d. 1.6 V; e. 0.4 V

EXPERIMENT 6-1

TIME-CONSTANT CIRCUITS

Inductance (L) opposes any change in current in a circuit, whereas capacitance (C) opposes any change in voltage. In both instances, the reaction time associated with inductance and capacitance in opposing changes to current and voltage depends on resistance. The time, in seconds, required for a capacitor or an inductor to react to change is determined with a quantity known as a time constant.

OBJECTIVE

To observe the time constant of a capacitor as the capacitor charges or discharges through a resistive path.

EQUIPMENT

Multimeter (VOM)

Variable dc power supply

Resistors: 220 kΩ, 100 kΩ

Capacitor: 47 μF

SPST switches (2) or DPDT (1)

Connecting wires

PROCEDURE

1. Construct the circuit in Fig. 6-1A.

2. When SPST switch S_1 is closed and SPST switch S_2 is open, the capacitor charges through R_1. After the capacitor is charged, it is discharged through R_2 by means of opening of switch S_1 and closing of switch S_2. Compute the time constant (t) for charging and discharging the capacitor in the circuit.

   Charging $t =$ _____ s.

   Discharging $t =$ _____ s.

3. Compute the total time required for the capacitor to charge to 25 V when S_1 is closed and S_2 is open. Total charge time = _____ s.

4. Compute the total time required for the capacitor to discharge 25 V when S_2 is open and S_2 is closed. Total discharge time = _____ s.

5. Compute the total charge and discharge currents for the circuit shown in Fig. 6-1A.

   Total charge current = _____ mA.

   Total discharge current = _____ mA.

6. Prepare the VOM to measure direct current in the 1 mA range and connect it in series with R_1 and C_1 at point A in the circuit shown in Fig. 6-1A. (*Note:* Both S_1 and S_2 must be open.)

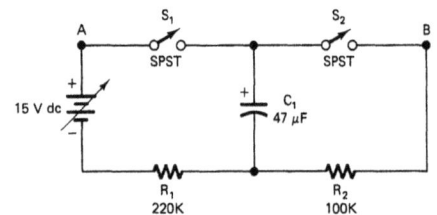

FIGURE 6-1A *RC* combination circuit.

7. Close S_1 and record the maximum charging current. Charging current = _____ mA.

8. Describe and explain the action of the circuit current as the capacitor is charged.

   _____

   _____

9. How did the computed charging current in step 5 compare with the measured charging current in step 7?

   _____

   _____

10. Disconnect the VOM, open S_1, and close S_2.

11. With the VOM prepared to measure the direct current described in step 6, connect this meter in series with C_1 and R_2 at point B in the circuit. (*Note:* Both S_1 and S_2 must be open.)

12. Close S_1 for 5 or 6 s. Open S_1, close S_2, and record the maximum discharge current. Discharge current = _____ mA.

13. Describe and explain the action of the current in the circuit as the capacitor discharged.

   _____

   _____

14. How did the computed discharge current in step 5 compare with the measured discharge current in step 12?

   _____

   _____

15. How did the charge and discharge currents of the capacitor compare? Why were they different?

   _____

   _____

16. Disconnect the VOM and open S_2.

17. Prepare the VOM to measure dc volts (25 V or higher range) and connect it across R_1 in the circuit.

18. Close S_1 and describe the voltage drop across R_1 as the capacitor charges.

   _____

   _____

19. Disconnect the VOM and place it across R_2.

20. Open S_1 and close S_2. Describe the voltage drop across R_2 as the capacitor discharges.

   _____

   _____

21. As the capacitor is charging, the current and the voltage across R_1 decrease. When the charging current and voltage across R_1 reach zero, the capacitor is fully charged. Likewise, as the capacitor discharges, the discharge current and voltage drop across R_2 decrease. When the discharge current and voltage across R_2 reach zero, the capacitor is fully discharged.

22. Disconnect the VOM and open S_1 and S_2.

23. Connect the VOM across R_1, as described in step 17.

24. Using a watch with a timer or a second hand, measure and record the time required for the capacitor to charge to 100% of the source voltage when S_1 is closed. This time corresponds to the time before the VOM will measure zero voltage across R_1. Charging time = _____ s.

25. How did the measured time in step 24 compare with the computed time in step 3?

   _____

   _____

26. Connect the negative lead of the VOM to the negative side of the capacitor. Do not connect the positive lead of the VOM into the circuit at this time.

27. Open S_1 and quickly connect the positive lead of the VOM to the positive side of the capacitor. Record the maximum voltage across the capacitor at the instant the positive lead is connected to the capacitor: Voltage across C = _____ V.

28. How did the voltage measured in step 27 compare with the source voltage?

   _____

   _____

ANALYSIS

1. What is the time constant of an RL circuit

   _____

   _____

2. What is the time constant of an RC circuit?

   _____

   _____

3. How many time constants are required for the current through an *RL* circuit to reach its maximum value? _____

4. How many time constants are required for the voltage across a capacitor in an *RC* circuit to equal the source voltage? _____

5. How can the length of an *RL* time constant be increased?

   _____

   _____

   Decreased?

   _____

   _____

6. How can the length of an *RC* time constant be increased?

   _____

   _____

   Decreased?

   _____

   _____

Unit 6 Examination

Inductance and Capacitance

Instructions: For each of the following, circle the answer that most correctly completes the statement.

1. Inductance is an electrical unit measured in
   a. Ohms   b. Farads
   c. Henrys   d. Volts

2. Four picofarads is equal to
   a. 0.004 F   b. 40 μF
   c. 4 μμF   d. 0.0004 F

3. Capacitance is measured in which of the following units?
   a. Farads   b. Henrys
   c. Ohms   d. Volts

4. A capacitor will charge to 95% of its full charge value after
   a. 1 time constant   b. 2 time constants
   c. 3 time constants   d. 4 time constants

5. A capacitor will discharge to 63% of its full charge value after
   a. 1 time constant   b. 2 time constants
   c. 3 time constants   d. 4 time constants

6. Counterelectromotive force is a property of which of the following devices?
   a. A resistor   b. A potentiometer
   c. A fuse   d. An inductor

7. Assume that a 0.0005 μF capacitor is fully charged 0.01 μs after power is applied to the capacitor. The value of the capacitor and the time required to charge the capacitor can also be expressed as
   a. 0.5 mF and 10 ms
   b. 500 pF and 10 ms
   c. 5.0 nF and 10 ns
   d. 500 pF and 10 ns

8. The additive formula $(C_T = C_1 + C_2 + C_3)$ is used for computing capacitance of parallel circuits. This additive relation exists because when capacitors are connected in parallel
   a. The distance between plates increases
   b. The plate area decreases

    c. The dielectric constant of the capacitors increases

    d. The area of plates increases

9. An inductor in a dc circuit will
    a. Oppose changes in current
    b. Oppose changes in voltage
    c. Stop all current flow
    d. Present an infinite resistance

10. *Capacitance* value is affected by
    a. Area of plates   b. Type of dielectric
    c. Distance between   d. All of the above
       plates

11. Which of the following is *not* a factor that affects the inductance (*L*) of a coil?
    a. Cross-sectional area of the coil
    b. Direction of current flow through the coil
    c. Number of turns of wire in the coil
    d. Type of core material used

12. Which of the following formulas would be used to calculate a correct *RC* time constant?
    a. $TC = R + C$ b. $TC = R \times C$
    c. $TC = R/C$ d. $TC = C/R$

True-False: Place either T or F in each blank.

_____ 13. The amount of source voltage controls the time constant of a circuit.

_____ 14. Increasing the resistance of an *RC* series circuit will affect the time constant.

_____ 15. Substituting a one microfarad capacitor for a 1 picofarad capacitor in a series *RC* circuit would increase the time constant.

_____ 16. The time constant of an *RL* circuit increases as the voltage across the resistor increases.

_____ 17. Current flows to and from the plates of a capacitor when the capacitor is charging.

_____ 18. In a complete *RC* series circuit with a dc source, the values of *R* and *C* determine the value of the source voltage.

_____ 19. The time constant of a series *RC* circuit is determined by means of multiplying the value of *R* in ohms by the value of *C* in farads.

_____ 20. A capacitor will charge to about 100% of its source voltage after four time constants.

Electronic Symbols

Fixed resistor		Ground		Fuse		
Tapped resistor		Contacts (normally closed)		Circuit breaker (single pole)		
Variable resistor (potentiometer)		Contacts (normally open)		Circuit breaker (three pole)		
Thermistor		Switch (single-pole, single-throw)		Coil (air core)		
Fixed capacitor		Switch (single-pole, double-throw)		Coil (iron core)		
Variable capacitor		Switch (double-pole, single-throw)		Coil (tapped)		
Polarized capacitor (electrolytic)		Switch (double-pole, double-throw)		Coil (adjustable)		

Battery		Multiposition selector switch (any number of positions may be shown.)		Transformer (air core)	
Alternating current source				Transformer (iron core)	
Piezoelectric crystal		Pushbutton switch (normally open)		Autotransformer	
Thermocouple		Pushbutton switch (normally closed)		Generator or motor field coil	
Thermal cutout device		Pushbutton switch (double circuit)		Antenna	or
Wires crossing; not connected		Limit switch (normally open)		Photovoltaic cell Solar cell	
Wires connected		Limit switch (normally closed)		Synchro unit	S_1 S_2 S_3 R_1 R_2
Female connector		Electrical bell		Meter	*
Male connector		Loudspeaker		*Replace with letter(s) designating type: V, A, MA, μA, W, etc.	
Joined connectors		Microphone	or	Generator	G or GEN
Jack (2-conductor)		Incandescent lamp		Motor	M or MOT
Plug (2-conductor)		Fluorescent lamp			

Semiconductor symbols:

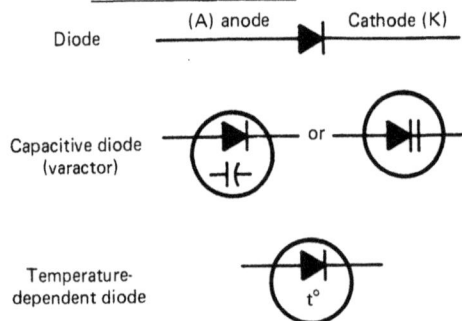

Diode	(A) anode	Cathode (K)	
Capacitive diode (varactor)		or	
Temperature-dependent diode	t°		

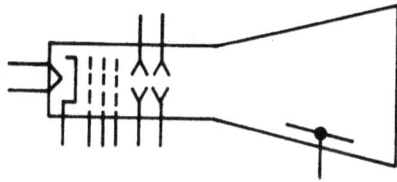

Cathode-ray tube
(electrostatic deflection)

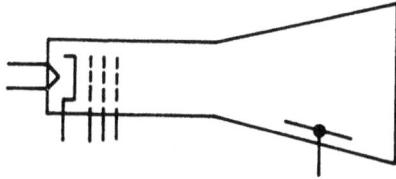

Cathode-ray tube
(electromagnetic deflection)

Thyristor,
bidirectional-
diode type

Thyristor,
bidirectional-
triode type (triac)

Bipolar
transistor
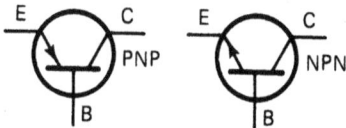

Photodiode
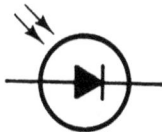

Light emitting
diode (LED)
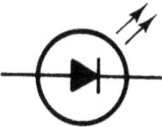

Zener
diode
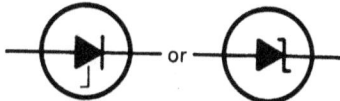

Thyrector
diode
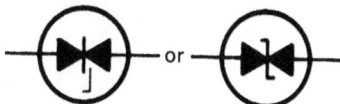

Tunnel
diode
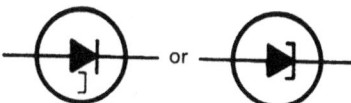

Trigger diac.
unidirectional
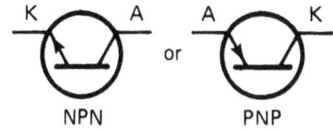

Thyristor,
reverse-
blocking-
diode type
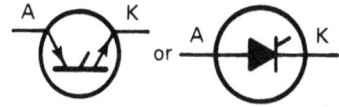

Thyristor,
reverse-blocking-
triode type
(solid-state thyratron, or SCR)

P-channel MOSFET,
enhancement type

Phototransistor
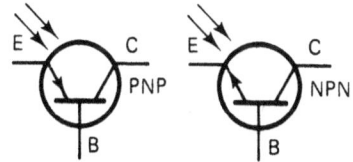

Darlington
transistor
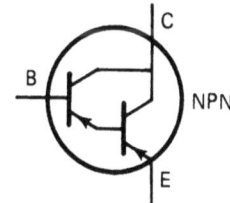

Unijunction
transistor
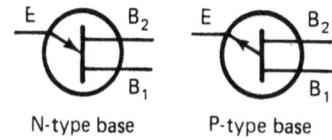

N-channel
JFET
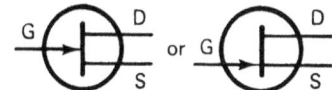

P-channel
JFET
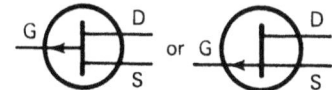

N-channel MOSFET, depletion type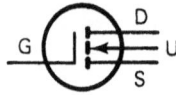

N-channel MOSFET, enhancement type

P-channel MOSFET, depletion type

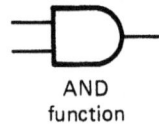

AND function

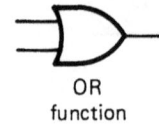

OR function

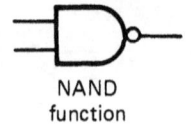

NAND function

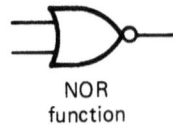

NOR function

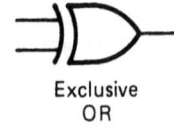

Exclusive OR function

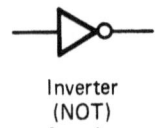

Inverter (NOT) function

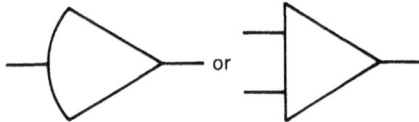

Operational amplifier

Flip-flop (general)

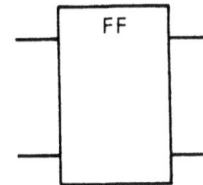

Electric Safety

Electric safety is very important. Many dangers are not easy to see. For this reason, safety should be based on understanding basic electric principles. Common sense also is important. Physical arrangement of equipment in the electric lab or work area should be done in a safe manner. Well-designed electric equipment always should be used. Electric equipment often is improvised for the sake of economy. It is important that all equipment be made as safe as possible. This is especially true for equipment and circuits that are designed and built in a lab or shop.

Work surfaces in the shop or lab should be covered with a material that is nonconducting. The floor of the lab or shop also should be nonconducting. Concrete floors should be covered with rubber tile or linoleum. A fire extinguisher that has a nonconducting agent should be placed in a convenient location. Extinguishers should be used with caution. The teacher should explain their use.

Electric circuits and equipment in the lab or shop should be plainly marked. Voltages at outlets require special plugs for each voltage. Several voltage values are used with electric lab work. Storage facilities for electric supplies and equipment should be neatly kept. Neatness encourages safety and helps keep equipment in good condition. Tools and small equipment should be maintained in good condition and stored in a tool panel or marked storage area. Tools that have insulated handles should be used. Tools and equipment plugged into convenience outlets should be wired with three-wire cords and plugs. The purpose of the third wire is to prevent electric shocks by means of grounding all metal parts connected to the outlet.

Soldering irons often are used in the electric shop or lab. They can be a fire hazard. They should have a metal storage rack. Irons should be unplugged while not in use. Soldering irons can cause burns if not used properly. Rosin-core solder always should be used in the electric lab or shop.

Adequate laboratory space is needed to reduce the possibility of accidents. Proper ventilation, heat, and light also provide a safe working environment. Wiring in the electric lab or shop should conform to specifications of the National Electric Code (NEC). The NEC governs all electric wiring in buildings.

Lab or Shop Practices

All activities should be conducted with low voltages whenever possible. Instructions for performing lab activities should be written clearly. All lab or shop work should emphasize safety. Experimental circuits always should be checked before they are plugged into a power source. Electric lab projects should be constructed to provide maximum safety when used.

Disconnect electric equipment from the source of power before working on it. For testing of electronic equipment, such as TV sets or other 120 V devices, an isolation transformer should be used. This isolates the chassis ground from the ground of the equipment and eliminates the shock hazard for work with 120 V equipment.

Electric Hazards

A good first-aid kit should be in every electric shop or lab. The phone number of an ambulance service or other medical services should be in the lab or work area in case of emergency. Any accident should be reported immediately to the proper school officials. Teachers should be proficient in the treatment of minor cuts and bruises. They also should be able to apply artificial respiration. In case of electric shock, when breathing stops, artificial respiration must be started immediately. Extreme care should be used in moving a shock victim from the circuit that caused the shock. An insulated material should be used so that someone else does not come in contact with the same voltage. It is not likely that a high-voltage shock will occur. However, students should know what to do in case of emergency.

The human body normally is not a good conductor of electricity. When wet skin comes in contact with an electric conductor, the body is a better conductor. A slight shock from an electric circuit should be a warning that something is wrong. Equipment that causes a shock should be immediately checked and repaired or replaced. Proper grounding is important in preventing shock.

Safety devices called *ground-fault circuit interrupters* (GFIs) are used for bathroom and outdoor power receptacles. They have the potential for saving many lives by preventing shock. GFIs immediately cut off power if a shock occurs. The National Electric Code specifies where GFIs should be used.

Electricity causes many fires each year. Electric wiring with too many appliances connected to a circuit overheats wires. Overheating may set fire to nearby combustible materials. Defective and worn equipment can allow electric conductors to touch one another and cause a short circuit, which causes a blown fuse. It also can cause a spark or arc, which might ignite insulation or other combustible materials or burn electric wires.

Fuses and Circuit Breakers

Fuses and circuit breakers are important safety devices. When a fuse blows, something is wrong in the circuit. The following could cause blown fuses:

1. A short circuit caused by two wires touching

2. Too much equipment on the same circuit

3. Worn insulation that allows bare wires to touch grounded metal objects, such as heat radiators or water pipes

After the problem is corrected, a new fuse of proper size should be installed. Power should be turned off to replace a fuse. Never use a makeshift device in place of a new fuse of the correct size. This destroys the purpose of the fuse. Fuses are used to cut off power and prevent overheating of wires.

Circuit breakers are now very common. Circuit breakers operate on spring tension. They can be turned on or off as wall switches can. If a circuit breaker opens, something is wrong in the circuit. Locate and correct the cause and then reset the breaker.

Always remember to use common sense whenever working with electric equipment or circuits. Safe practices should be followed in the electric lab or shop and in the home. Detailed safety information is available from the National Safety Council and other organizations. It is always wise to be safe.

Electronic Equipment and Parts Sales

The following are the names and addresses of several companies that sell electronic equipment and parts. You can write to these companies and obtain catalogs and price lists for purchasing the equipment you need.

All Electronics Corp.
905 S. Vermont Ave.
P.O. Box 20406
Los Angeles, CA 90006
800-826-5432

Allied Electronics
1355 N. McLean Blvd.
Elgin, IL 60120
800-433-5700

Brodhead-Garrett
4560 E. 71st St.
Cleveland, OH 44105
216-341-0248

Cal West Supply, Inc.
31320 Via Colinas, Suite 105
Westlake Village, CA 91362
800-892-8000

Circuit Specialist Co.
P.O. Box 3047
Scottsdale, AZ 85257

Digi-Key Corp.
P.O. Box 677
Thief River Falls, MN 56701

Edlie Electronics
2700 Hempstead Twp.
Levittown, NY 11756-1443
800-645-4722

ETCO Electronics
North Country Shopping Ctr.
Plattsburgh, NY 12901

Hewlett-Packard
1501G Page Mill Rd.
Palo Alto, CA 94305

Hickok Teaching Systems
2 Wheeling Ave.
Woburn, MA 01801
617-935-5850

Hughes-Peters
4865 Duck Creek Rd.
P.O. Box 27119
Cincinnati, OH 45227
800-543-4483

Jameco Electronics
1355 Shoreway Rd.
Belmont, CA 94002

Kelvin Electronics, Inc.
P.O. Box 8
1900 New Hwy.
Farmingdale, NY 11735
800-645-9212

Lab Volt
Buck Engineering Co.
Farmingdale, NJ 07727

MCM Electronics
858 E. Congress Park Dr.
Centerville, OH 45459

Merlin P. Jones & Assoc.
PO Box 12685
Lake Park, FL 33403-0685
305-848-8236

Mouser Electronics
2401 Hwy. 287 N
Mansfield, TX 76063

Omnitron Electronics
770 Amsterdam Ave.
New York, NY 10025
800-223-0826

Priority One Electronics
21622 Plumer St.
Chatsworth, CA 91311
800-423-5922

RNJ Electronics
805 Albany Ave.
Lindenhurst, NY 11757
800-645-5833

Satco
924 S. 19th Ave.
Minneapolis, MN 55404
800-328-4644

Techni-Tool
P.O. Box 368
Plymouth Meeting, PA 1946
215-825-4990

Tektronix, Inc.
P.O. Box 1700
Beaverton, OR 97075

URI Electronics
P.I. Burks Co.
842 S. 7th St.
Louisville, KY 40203

Soldering Techniques

Soldering is an important skill for electric technicians. Good soldering is important for the proper operation of equipment.

Solder is an alloy of tin and lead. The solder used most is 60/40 solder. This means that it is made from 60% tin and 40% lead. Solder melts at a temperature of about 400°F.

For solder to adhere to a joint, the parts must be hot enough to melt the solder. The parts must be kept clean to allow the solder to flow evenly. Rosin flux is contained inside the solder. It is called *rosin-core solder*.

A good mechanical joint must be made when soldering. Heat is then applied until the materials are hot. When the materials to be soldered are hot, solder is applied to the joint. The heat of the metal parts (not the soldering tool) is used to melt the solder. Only a small amount of heat should be used. Solder should be used sparingly. The joint should appear smooth and shiny. If it does not, it could be a "cold" solder joint. Be careful not to move the parts when the joint is cooling. This could cause a "cold" joint.

When parts that can be damaged by heat are soldered, be very careful not to overheat them. Semiconductor components, such as diodes and transistors, are highly heat sensitive. One way to prevent heat damage is to use a *heat sink*, such as a pair of pliers. A heat sink is clamped to a wire between the joint and the device being soldered. A heat sink absorbs heat and protects delicate devices. Printed circuit boards also are highly sensitive to heat. Care should be taken not to damage printed circuit boards when soldering parts onto them.

Several types of soldering irons and soldering guns are available. Small, low-wattage irons should be used with printed circuit boards and semiconductor devices.

The following are some rules for good soldering:

1. Be sure that the tip of the soldering iron is clean and tinned.

2. Be sure that all the parts to be soldered are heated. Place the tip of the soldering iron so that the wires and the soldering terminal are heated evenly.

3. Be sure not to overheat the parts.

4. Do not melt the solder onto the joint. Let the solder flow onto the joint.

5. Use the right kind and size of solder and soldering tools.

6. Use the right amount of solder to do the job but not enough to leave a blob.

7. Be sure not to allow parts to move before the solder joint cools.

Troubleshooting

Troubleshooting is a method of finding out why something does not work properly. If you follow logical steps, you will be able to locate most difficulties that occur in electronic equipment. Sometimes the trouble is so complex that it requires many hours of concentration and work. Other problems are easy to solve and require only a brief time.

Resources that most people find helpful in troubleshooting are as follows:

1. Using a common sense approach
2. Knowing how electronic systems work
3. Knowing how to use test equipment
4. Knowing how to use schematics effectively
5. Being able to find the trouble through a logical sequence

To begin any kind of troubleshooting, you should first determine possible courses of action. Without a system, the procedure of troubleshooting becomes a guessing game. You should be aware that no one system of troubleshooting is perfect. In the process of troubleshooting, keep in mind that most problems are component failures. If you know what each component is supposed to do, you will be aware of the troubles they can cause.

During troubleshooting, it is important that you use proper tests. Much of your time is used in locating the trouble. You must have a suitable approach to save time. As you become more familiar with troubleshooting, it becomes less time consuming.

As you continue your troubleshooting effort, you must constantly keep aware of circuit or system operations you have already tested. Make a list or remember the probable troubles that have been tested.

An important part of troubleshooting skills is the initial inspection. Initial inspection involves looking for the obvious. You should do several things before you perform actual circuit or system testing. In the initial inspection of any equipment, first open the equipment to look at it. There are several things you should observe, as follows:

1. Burned resistors. They are often obvious, may be charred, blistered or bulged, or have discolored color bands and even holes.
2. Broken parts. These may come in the form of cracks, wires pulled out of parts, or destroyed parts.
3. Broken wires and poor connections.
4. Smoke or heat damage. Parts may smoke when equipment is turned on; this identifies defective parts (but not the cause).
5. Oil leaks and water leaks.
6. Loose, damaged, or worn parts. These are determined by means of visual and tactile examination.
7. Noisy parts. Uncommon noises indicate defective parts.

When initial inspection is performed properly, many troubles can be located without having to go through unnecessary steps. Initial inspection involves the senses of sight, touch, smell, and hearing. It is important to organize your thoughts to solve the problem. If you suspect that a part is the source of the problem, take a closer look. If you suspect a specific part, turn off the equipment. Smell it, touch it, and examine it closely. For example, a transformer that is good does not have an odor, but a burned transformer does. Initial inspection can help to locate the trouble in any defective electronic equipment or circuits.

The final solution of the problem involves application of your *knowledge of electronic circuit operation and understanding of proper use of test equipment.* Remember that troubleshooting is a systematic procedure.

Index